OPTOELECTRONIC DEVICES AND OPTICAL IMAGING TECHNIQUES

DOUGLAS A. ROSS

Queen Mary College,
London

M

First published 1979 by
THE MACMILLAN PRESS LTD
London and Basingstoke
Associated companies in Delhi Dublin
Hong Kong Johannesburg Lagos Melbourne
New York Singapore and Tokyo

Typeset in 10/12 Press Roman by
Styleset Limited, Salisbury, Wilts.
and printed in Great Britain by
Unwin Brothers Limited
The Gresham Press
Old Woking, Surrey

British Library Cataloguing in Publication Data

Ross, Douglas A
 Optoelectronic devices and optical imaging
 techniques.
 1. Electrooptical devices
 I. Title
 621.36   TA1750

 ISBN 0–333–25334–5
 ISBN 0–333–25335–3 Pbk

This book is dedicated to Milton Paul Daniels

OPTOELECTRONIC DEVICES AND
OPTICAL IMAGING TECHNIQUES

Contents

Preface

The technology of optoelectronics, which deals with devices combining optical and electrical ports, has expanded enormously in the past few years, as a result of the rapid expansion of the semiconductor industry. Some optoelectronic devices, such as the light emitting diode or LED, have found their way into the majority of electronic appliances manufactured today. This rather radical change from tungsten filament light sources, producing more heat than light, to LEDs producing highly coloured or monochromatic light, has taken place almost without comment.

The fact that the present age could be described as the era of the crystalline semiconductor is reflected in the variety of optoelectronic devices. Almost every function which formerly required expensive photosensitive or photoemissive devices operating at high voltages, can now be accomplished with an inexpensive semiconductor device requiring a few volts. The photomultiplier tube can be replaced by the photodiode/operational amplifier; the gas laser by the laser diode; the cathode ray oscilloscope by the LED display; the television camera by the charge coupled device area sensor; and they can all be powered by another optoelectronic device, the solar cell.

This book is written in a spirit of excitement and wonder at the ingenuity of man to develop new solutions to old problems. It is hoped that the reader may sense some of the author's enthusiasm, and may even be carried away to develop his own interest in optoelectronics. The book is written at a level which should be accessible to the undergraduate student with some understanding of semiconductor physics and electronic circuits. The more difficult topics in the book are included for completeness and for the reader who wants to develop more than a superficial understanding.

The devices described here reflect the author's interests and are not meant to provide an encyclopedic coverage of the subject. None the less, it is hoped that a representative collection of devices has been included, and that these will maintain their importance in the future. It is the task of the present generation of engineering and physics students to solve some of the many problems hinted at here; this is, after all, only a beginning.

London, May 1979   DOUGLAS A. ROSS

1 Photons and Matter

1.1 Introduction

Optoelectronic devices are those which convert light into electrical energy or vice versa. In some cases the same device may perform both functions. In the past few years, a host of new devices has been developed for commercial applications – LEDs, photodiodes, solar cells, laser diodes, and many more.

In this book several optoelectronic devices are discussed in some detail. It is hoped that this will serve as a useful introduction to those engineers and physicists who are curious about this rapidly developing field. In this introductory chapter a few underlying properties of photons and semiconductors are discussed and these will be referred to again and again throughout the book. The following chapters deal in turn with the LED, the photoconductor or light sensitive resistor, the photodiode and phototransistor, the solar cell, and the laser diode. The final chapter of the book is a discussion of optical imaging techniques.

1.2 Photons

The nature of light has been a topic of controversy in physics for centuries. The debate has evolved into two schools of thought – the wave theory of light and the particle theory of light. Both points of view can be supported by experimental evidence. Light is diffracted by edges or by periodic arrays such as the diffraction grating, as are electromagnetic waves. On the other hand, it can be shown that the photocurrent generated by a photomultiplier tube is composed of individual pulses, each representing the arrival of one particle of light. The wave theory of light gained enormous prestige from Maxwell's theory of electromagnetic waves, when it was shown that visible light was an electromagnetic phenomena. However, it was not long before the particle theory of light was revived by Planck's introduction of the concept of quantisation of energy, in order to explain the blackbody radiation spectrum, Planck's quanta soon led to an entirely new physics, quantum mechanics.

The present view in physics is that we must accept both the wave and the particle properties of light. We imagine that light is composed of discrete lumps of energy, photons, which behave both like waves and particles. It turns out that in many situations there are such a large number of photons that the wave-like behaviour is dominant. For instance, a 1 mW He–Ne gas laser emits of the order of 10^{15} photons per second, and most of the characteristics of the emitted beam of light can be explained by the theory of plane waves.

We are familiar with the wavelengths of different photons through the phenomenon of colour vision. White light may be diffracted into its component colours by a prism or by a diffraction grating. The ratio of the spacing between lines of the grating and the wavelength determines the angle of diffraction of each colour.

The photon travels at the speed of light, $c = 2.998 \times 10^8$ m/s in a vacuum, and it is pure energy having zero rest mass. It can be thought of as a packet of waves whose energy and frequency are related by $E_0 = hf$. The wavelength of this wave packet is given by $\lambda = c/f$. This can be seen by imagining an oscillation at frequency f which travels with speed c, as in figure 1.1. The distance between successive peaks of the oscillation is the wavelength λ. Note that photons representing different colours of light have different energy, and that those of green light carry more energy than those of red light.

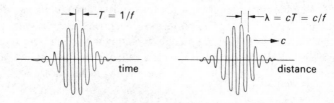

FIGURE 1.1

An artist's impression of a photon of frequency f moving at speed c

Most of this book is concerned with interactions between photons and matter; the interface between the two is the electron, and when a photon is absorbed in a semiconductor, an electron gains its energy and is momentarily free to conduct. When a free electron becomes trapped it loses energy, which may be released by the emission of a photon. The photon may seem an elusive concept, but its reality must be accepted in order to understand optoelectronic devices, which are all based on the photon—electron interaction. In the next section, some general properties of matter, beginning with the hydrogen atom, are discussed.

1.3   Energy Diagrams

The properties of the hydrogen atom have been studied extensively in quantum mechanics, since in the case of hydrogen — one proton orbited by one electron — it is possible to solve Schrödinger's equation. It is found that the electron can have only one of certain discrete energies. These are classified so that the lowest, −13.6 eV, is the amount of energy required to ionise hydrogen in its ground state, that is, to remove its electron to infinity. The electron volt (eV) is the

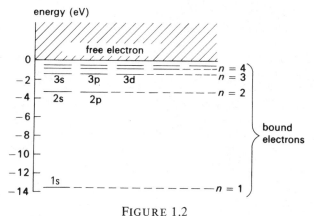

FIGURE 1.2

The energy diagram of a hydrogen atom

amount of energy gained by an electron accelerated through a potential of 1 V, 1.602×10^{-19} joules (J).

Bound electrons in atoms have negative energies, since they cannot escape unless acted upon by external forces which contribute a net positive energy equal to, or greater than, the ionisation energy. The various possible energies that an electron in hydrogen can have are shown in figure 1.2. At room temperature the hydrogen atom may not be in its ground state. For instance, hydrogen gas at room temperature has many atoms in the ground state, some in the 2s state, some in the 2p state, and so on. If a particular atom has an electron in the 3s state, for example, then it usually gives up some of its energy in assuming a lower energy state almost immediately. When this happens, the energy loss of the electron is released as a photon, having energy equal to the energy transition. In a transition from the 3s to the 1s state, for instance, the emitted photon would have energy $hf = -1.51$ eV $- (-13.6$ eV$) = 12.1$ eV. Here f is frequency (in Hz) of the photon and $h = 6.626 \times 10^{-34}$ J s is Planck's constant. The photon's frequency may be calculated by converting 12.1 eV into joules and dividing by Planck's constant. The result is $f = 2.92 \times 10^{15}$ Hz.

It is possible to determine the energy diagram of the hydrogen atom experimentally by studying the spectrum of emitted radiation from hydrogen gas. Each energy transition will yield photons of a given energy, or frequency (see table 1.1). Since all photons move at the speed of light, $c = 2.998 \times 10^8$ m/s in a vacuum, they will have a wavelength given by $\lambda = c/f$. In the case of the 3s to 1s transition, the emitted photons will all have a wavelength equal to 102.6 nm (1 nm $= 10^{-9}$ m) or 1026 Å (1 Å $= 10^{-10}$ m). Thus, the spectrum of radiation emitted from hydrogen gas will have a series of peaks at wavelengths representing the different possible energy transitions.

By using the energy diagram for hydrogen, it is possible to build up the electron configuration of all the atoms in the periodic table. The only modifica-

TABLE 1.1

Basic energy transitions of the hydrogen atom, and the frequency and wavelength of the corresponding emitted photon

Transition		Emitted photon		
Energy transition	ΔE (eV)	Frequency (Hz)	Wavelength (nm)	
5–1	13.1	3.16×10^{15}	94.9	
4–1	12.8	3.08×10^{15}	97.2	ultra-
3–1	12.1	2.92×10^{15}	103.0	violet
2–1	10.2	2.47×10^{15}	122.0	
5–2	2.86	6.91×10^{14}	434.0	
4–2	2.55	6.17×10^{14}	486.0	visible
3–2	1.89	4.57×10^{14}	656.0	
5–3	0.97	2.35×10^{14}	1280.0	
4–3	0.66	1.60×10^{14}	1880.0	infra-
5–4	0.31	0.75×10^{14}	4000.0	red

tion is that because the higher elements have a nucleus with several protons, the electrostatic force between the electrons and nucleus is generally greater than in hydrogen. For instance, the ground state of helium has two electrons in the 1s state, with an ionisation energy of −24.6 eV. It takes more energy to ionise helium than hydrogen, which explains why helium is more stable.

Although the Pauli exclusion principle specifies that two electrons may not occupy the same state, the two electrons of the helium atom can share the 1s state because they have opposite spin. In the simplified energy diagram of figure 1.2, we have neglected the splitting of the various energy levels into sub-levels which results from electron spin.

One of the most common materials used as a semiconductor is silicon, whose atoms have 14 protons in their nucleus. Silicon has two electrons in the 1s state, two in the 2s state, six in the 2p state (two electrons in each of three states of angular momentum), two in the 3s state, and two in the 3p state. The energy of the most loosely bound electron of silicon is −8.1 eV. The four electrons in the 3s and 3p states all have about this energy, consequently silicon is said to have a valence of 4 since it can share these four electrons with other atoms relatively easily. In a semiconductor crystal formed of silicon, the atoms form a tetrahedral pattern, each atom surrounded by four equidistant nearest neighbours. This pattern is characteristic of diamond, silicon, germanium and any crystal made up of atoms with four electrons which are relatively loosely bound.

A semiconductor crystal has an energy diagram quite different from that of an atom of the material on its own. If we imagine a very large number of atoms in close proximity in the crystal, they are perturbed by the fluctuating thermal energy of the crystal and by the electrostatic forces they exert on each other. These perturbations cause further splitting of the energy levels, and since there are very many atoms in a typical crystal, these very closely spaced but discrete energy levels blend together to form energy bands.

In crystalline silicon there exists an energy band, called the valence band, which represents the energy of bound electrons in the 3s and 3p states of the silicon atoms, and a higher energy band, called the conduction band, which represents the energy of electrons which are free to move about in the crystal. The energy gap between the valence and conduction bands in crystalline silicon at room temperature is 1.12 eV, similar in magnitude to the energy difference between the 3s and 4s levels of a silicon atom.

At room temperature there is enough thermal energy to ionise some of the atoms of a semiconductor material, and the material has a small but significant conductivity due to a small population of electrons in its conduction band, hence the name semiconductor. If a particular electron gains enough energy to escape its host atom, it executes a random walk through the material until encountering an ionised atom, or hole, where it recombines. In the process of recombination it loses energy equal to the energy gap between valence and conduction bands, and can release this energy in the form of a photon. The photon will have a frequency given by $f = E_g/h$, where E_g is the energy gap of the material. In silicon $E_g = 1.1$ eV and $f = 2.66 \times 10^{14}$ Hz ($\lambda = 1127$ nm).

One point which should be mentioned about energy diagrams for semi-conductors is that free electrons have momentum as well as energy. An electron which has an energy E_0 at rest has total energy $E_0 + \frac{1}{2}mv^2$ when moving with speed v. The momentum of a moving electron is $p = mv$. Thus its total energy is $E_0 + p^2/2m$. An energy diagram of such an electron is a plot of its total energy versus momentum, as in figure 1.3. Note in this figure that if the total energy is E, then $\partial E/\partial p = p/m$, and $\partial^2 E/\partial p^2 = m^{-1}$. Thus the mass of the electron may be computed from the inverse of the inflection of the energy diagram $m = (\partial^2 E/\partial p^2)^{-1}$.

Although the electrons of a semiconductor are not totally free, they do enjoy a certain mobility since they can move through the material until recombining with a hole. In the same way a hole has mobility in a semiconductor, since the site of ionisation of atoms moves about randomly. In the energy diagram of a semiconductor, the momentum of both electrons and holes must be considered. This is illustrated in figure 1.4, which shows the energy diagrams of silicon, germanium and gallium arsenide.

In this figure we could imagine that an electron which is in the conduction band, that is, one which has escaped from its host atom, behaves like a free particle with effective mass $m^* = (\partial^2 E/\partial p^2)^{-1}$. In the same way the effective mass of holes in the material may be computed from the inflection of the curve for the valence band.

In future chapters we shall almost always refer to the type of simplified

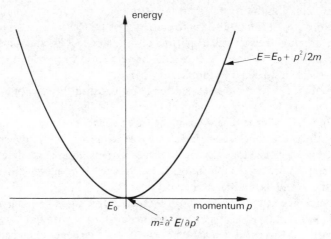

FIGURE 1.3
The energy diagram of a free electron

energy diagram as in figure 1.5, but the reader should remember the true situation. Before leaving this subject note that both silicon and germanium are indirect gap materials, that is, the bottom of the conduction band is at a different momentum to the top of the valence band. Consequently, when an electron and hole recombine in silicon or germanium, the electron must lose momentum as well as energy, and this loss of momentum goes into vibration of the crystal lattice of the material. For this reason, neither silicon nor germanium are efficient emitters of photons. On the other hand, GaAs is a direct gap material, and emits photons readily. It is one of the primary electroluminescent semi-conductor materials. Silicon and germanium find a role as absorbers of photons in photodiodes and photoconductors.

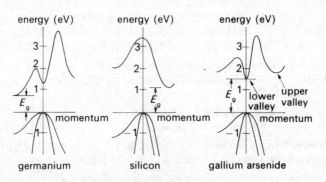

FIGURE 1.4
The energy diagrams of silicon, germanium and gallium arsenide

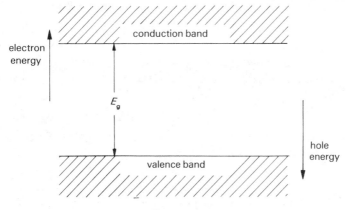

FIGURE 1.5
A simplified energy diagram

1.4   Charge Mobility

When a voltage is applied to the terminals of a semiconductor, a current flows. This is an external description of the somewhat complicated phenomenon of charge mobility. The applied voltage produces an electric field in the interior of the semiconductor, which causes charges to move under the influence of the force $F = \pm eE$. Positive charges, or holes, move in the direction of the applied field and the electrons move in the opposite direction. The external current is just the product of the total charge passing through a cross-section of the material x the average velocity of the charges. If both holes and electrons are moving, there is a contribution to the external current from each. Usually the electrons move more rapidly than do the holes.

The charges do not move in such a regular way as this description might imply. A semiconductor at room temperature has enough thermal energy for there to be a very large number of free electrons, and a correspondingly large number of holes. A hole is just an atom of the semiconductor crystal which has lost one of its electrons. Although we think of holes as having a certain mobility, this means that the site of ionisation changes, the atoms are fixed in position. The electrons, which can move freely in the semiconductor crystal lattice, do not move very far in any one direction before colliding with a hole. These collisions either cause the electron to be deflected away, or else the electron and hole combine, and the electron is said to have passed into the valence band. The path of an electron in a semiconductor could be described as a random walk, which, if there is an externally applied electric field, has a bias in a particular direction (see figure 1.6).

If an electron was free to move without collisions, it would gain more and more speed as a result of the constant acceleration of the applied electric field.

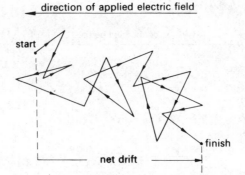

FIGURE 1.6

The random walk of an electron through a semiconductor crystal with applied external voltage

At each collision the electron gives up some of the energy it has gained, and consequently it never achieves very high velocities. The net result is that the externally applied field causes the electron to drift in a particular direction, even though its motion is constantly interrupted by collisions. The electron mobility in a semiconductor is defined by

$$v_n = \mu_n E \qquad \text{(electron mobility)} \qquad (1.1)$$

where v_n is the drift velocity, E the applied electric field, and μ_n is the electron mobility. Note that if the voltage is doubled (E doubled) the current is doubled (v_n doubled) in accordance with Ohm's law.

If the drift velocity has units of cm/s and the applied electric field units of V/cm, the electron mobility has units of cm^2/V s. In silicon the electron mobility has a value of around 1500 cm^2/V s. This means that 1 V applied over 1 cm produces an electron drift velocity of 1500 cm/s.

The hole mobility in a semiconductor is defined by

$$v_p = \mu_p E \qquad \text{(hole mobility)} \qquad (1.2)$$

In silicon μ_p has a value of around 600 cm^2/V s, about one-third the electron mobility.

In a highly doped semiconductor, which has more impurity atoms, there are more sites of ionisation, and hence the electrons undergo more collisions. The effect of this is to reduce the mobility. For example, in silicon the electron mobility drops from 1500 cm^2/V s to 700 cm^2/V s if the impurity concentration is increased by three orders of magnitude (10^{14} to 10^{17} per cm^3). The mobility is also reduced at higher temperatures since the increased thermal energy causes a greater proportion of ionised atoms, and hence more collisions.

If a free electron in silicon is accelerated by an electric field strength of 1V over 100 μm, or $E = 100$ V/cm, then it experiences an acceleration of $eE/m^* =$

1.76×10^{15} cm/s^2, assuming an effective mass $m^* = 9.106 \times 10^{-31}$ kg. If it is accelerated for 1 ns before a collision, it gains a speed of 1.76×10^6 cm/s and travels a total distance of 8.8 μm (1 μm $= 10^{-4}$ cm), passing through 16 000 atoms. It is rather unlikely that a free electron could survive for such a long period of time between collisions. The lattice constant for crystalline silicon, which defines the dimensions of one 'cell' of its crystal structure, is 0.543 nm (1 nm $= 10^{-7}$ cm). None the less it is easy to see that the drift velocity is increased by fewer collisions, and that in the detailed motion of the electron its instantaneous speed may be orders of magnitude larger than its drift velocity.

1.5 Absorption

When light passes through a thin layer of semiconductor, a proportion is absorbed and the remainder transmitted through. In terms of photon–electron interactions, a proportion of the incident photons are converted into conduction or 'free' electrons, and the remainder pass straight through the material without being affected. The basic equation of absorption is

$$I(x) = I_0 \exp(-\alpha x)$$

where I_0 is the incident light intensity, $I(x)$ the intensity at a distance x into the semiconductor, and α is the absorption coefficient.

Since we know that only those photons with enough energy to overcome the energy gap between valence and conduction bands of the semiconductor are

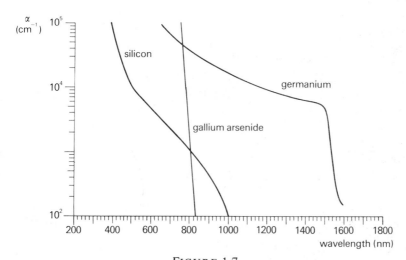

FIGURE 1.7

The variation of the absorption coefficients of silicon, germanium and gallium arsenide with wavelength

absorbed, it is very likely that the absorption coefficient will be highly wavelength dependent. This is certainly the case in silicon, germanium and gallium arsenide, as may be seen in figure 1.7.

The effective depth of penetration of light into a semiconductor is the distance at which its intensity is reduced to $1/e$ of its initial value. Thus the effective depth of penetration is α^{-1}, the inverse of the absorption constant. For example, silicon has an absorption constant of $\alpha = 4000$ cm^{-1} at a wavelength of 632.8 nm, and an effective depth of penetration of 2.5 μm.

The properties of the three semiconductors germanium (Ge), silicon (Si) and gallium arsenide (GaAs) are summarised in table 1.2. These are the three most widely studied semiconductor materials, although others may play an equally important role in the future. This information is intended only for reference in later chapters. The reader should consult a text on the physics of semiconductor devices for a complete discussion of these and other properties of semiconductors. We shall attempt to explain the more important properties as they arise throughout the book, and it is hoped that the reader may be able to follow the subject matter with a minimal amount of additional reading.

TABLE 1.2

Properties of Ge, Si and GaAs at room temperature (T = 300 K)

Properties	Ge	Si	GaAs
Atoms/cm^3	4.42×10^{22}	5.0×10^{22}	2.21×10^{22}
Atomic weight	72.6	28.08	144.63
Voltage breakdown (V/cm)	10^5	3×10^5	4×10^5
Crystal structure	diamond	diamond	zincblende
Density (g/cm^3)	5.3267	2.328	5.32
Dielectric constant	$16\epsilon_0$	$11.8\epsilon_0$	$10.9\epsilon_0$
Density of states in conduction band	1.04×10^{19}	2.8×10^{19}	4.7×10^{17}
Density of states in valence band	6.1×10^{18}	1.02×10^{19}	7.0×10^{18}
Energy gap (eV)	0.66	1.12	1.43
Intrinsic carrier concentration (cm^{-3})	2.4×10^{13}	1.6×10^{10}	1.1×10^7
Lattice constant (Å)	5.65748	5.43086	5.6534
Melting point ($^\circ$C)	937	1420	1238
Minority carrier lifetime (s)	10^{-3}	2.5×10^{-3}	10^{-8}
Mobility (cm^2/V s)			
μ_n (electrons)	3900	1500	8500
μ_p (holes)	1900	600	400

From S. M. Sze, *Physics of Semiconductor Devices* (John Wiley, London, 1969).

2 The Light Emitting Diode

2.1   Electroluminescent Devices

The sources of optical radiation between approximately 0.1 μm and 1 μm wavelength (1 μm = 10^{-6} m) may be classed into two types according to their spectral line width. The communication engineer is aware of the desirability of a signal source which is monochromatic. Such a single frequency source may be modulated and can act as the carrier of a high concentration of information. On the other hand, a wide bandwidth source would not be suitable for communication applications, since the modulation signal would become irretrievably scrambled in the transmission and detection process. Thus, the most important question we may ask about an optical source is whether or not its emission has a narrow spectral width.* The coherent sources, the gas, liquid or solid state lasers, have a spectral width of the order of 0.01 to 0.1 nm giving a relative bandwidth of the order of 10^{-5} to 10^{-4} (the line width of a He–Ne laser emitting red light at λ_0 = 632.8 nm is Δf = 7.5 GHz). The incoherent sources, the electroluminescent devices such as the light emitting diode, have a spectral width of the order of 10 nm with a relative bandwidth of the order of 10^{-2}. The applications of electroluminscent devices are mainly in optical coupling, and optical display and illumination.

Electroluminescence is the emission of optical radiation as a result of excitation by electric field or current. The transitions which may result in luminescent emission are summarised in figure 2.1. The first three represent transitions of electrons from the conduction band to an impurity energy above the valence band, or from the impurity level below the conduction band to the valence band. These donor or acceptor impurities result from doping or from structural defects in the material. In all three cases the transition energy is less than the energy gap of the material ($hf < E_g$). The next three represent band-to-band transitions which may exceed the energy gap of the material ($hf > E_g$). This occurs when energetic electrons are introduced into the conduction band of the material. Not all of these transitions occur in the same material, nor do all energy transitions release photons. Efficient luminescent materials are those in which radiative transitions predominate.

Radiative transitions in luminescent materials can occur over a range of transition energies centred around the band gap of the material. This accounts for the large spectral width of an electroluminescent device, the higher energy

---

* The spectral width of a source is the width of its emission spectrum, expressed in terms of wavelength, measured between half power points of the spectrum.

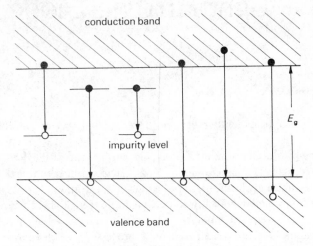

FIGURE 2.1

Various energy transitions in a doped semiconductor

transitions giving photons with a shorter wavelength, while longer wavelength photons result from the lower energy transitions involving trapping levels. The relationship between the energy gap of the material and the corresponding emission frequency or wavelength is

$$E_g = hf = \frac{hc}{\lambda}$$

where $h = 6.626 \times 10^{-34}$ J/s, and is Planck's constant, f is the frequency of emitted photon, λ is the wavelength of emitted photon, and $c = 2.998 \times 10^8$ m/s, which is the speed of light in vacuum. Usually the gap energy is expressed in electron volts (eV) rather than joules (J).

$$1 \text{ eV} = 1.602 \times 10^{-19} \text{ J}$$

$$\lambda \text{ (nm)} = \frac{1240}{E_g(\text{eV})}$$

$$f \text{ (THz)} = 242 \, E_g \text{ (eV)}$$

$$1 \text{ THz} = 10^{12} \text{ Hz}$$

Although it is not usual to compute the frequency of an optical signal, it is helpful to remember that a photon is a burst of electromagnetic energy at a frequency of a few hundred terahertz, or a wavelength of a few hundred nanometers.

In the beginning of this chapter it was mentioned that the wide spectral width of electroluminescent devices makes them unsuitable for communication applications. This is not precisely correct, but the wide spectral width severely limits the bandwidth of modulation signal which may be accommodated. For example, the light emitting diode (LED) has a very fast turn on time, of the order of 1 ns = 10^{-9} s. Suppose information is sent by means of pulse code modulation (PCM). Thus a 1 ns burst of red light from the LED would constitute a single pulse. If this pulse is sent through the atmosphere to a photodetector we have an optical communication link. However, atmospheric absorption of visible and infrared light limits the range of such a system to a mile or less. This problem may be overcome by the use of optical waveguide made from extremely fine glass fibre, drawn to 1 km lengths. The absorption of optical fibre has been reduced to acceptable levels to make possible optical communication over long distances. Having overcome the problem of absorption of the LED output, a much more subtle problem arises, dispersion! The index of refraction of glass, and hence glass fibre, varies with wavelength, as any schoolboy knows who has used a prism to split white light into its rainbow spectrum. Thus the speed of light travelling in an optical fibre varies with wavelength. This means that the different wavelengths in a 1 ns pulse (with 10 nm spectral width) will travel at different speeds to the photodetector, causing broadening of the pulse. Generally speaking, the wider the spectral width of the source, the more broadening will result when pulses are transmitted. Because of dispersion the pulse rate of the LED output must be limited to prevent pulses merging into one another, and this limits the bandwidth of modulation signal which may be sent. It is likely that the laser diode, with a spectral width of the order of 0.1 nm, will be the source used in optical fibre communication links.

Electroluminescent devices find application where large signal bandwidth is not required, and as lamps, digital displays and optical couplers for electrical isolation of circuits.

2.2 The Light Emitting Diode (LED)

A typical light emitting diode consists of a GaAsP (gallium arsenide phosphide) epitaxial layer grown on a GaP substrate, as shown in figure 2.2. The transparent GaP substrate allows a more efficient light output, since even the light which is emitted downwards into the substrate is reflected upwards and passes out of the device. A p-type layer is diffused on to the n-type GaAsP layer, producing a p–n junction as shown.

A high efficiency of light emission is obtained from this device since GaAsP acts as a direct gap material. If the LED is forward biased then the forward current which results causes the injection of conduction electrons into the junction region. When these conduction electrons recombine, energy is released in the form of photons with wavelength λ (nm) = $1240/E_g$ (eV), where E_g

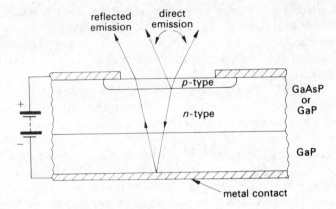

FIGURE 2.2

GaAsP or GaP on transparent GaP substrate

depends on the doping. For emission in the red (λ = 635 nm) a gap energy of approximately 1.95 eV is required.

Over a range of currents the number of photons emitted per unit time is proportional to the number of electrons injected into the conduction band of the diode per unit time. Thus the output light intensity is proportional to input current for forward current up to a few tens of milliamperes. A further increase in current causes saturation of the LED output light intensity (see figure 2.3). This occurs because the number of electrons being injected into the conduction band per unit time exceeds the transition rate of electrons passing from

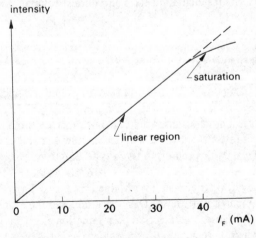

FIGURE 2.3

LED response curve

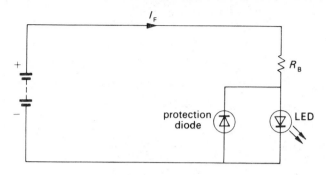

FIGURE 2.4
A simple LED bias circuit

conduction to valence band. Also at higher currents the power dissipation of the LED may exceed maximum levels and destroy the device.

In a typical bias circuit (for example, figure 2.4) a series resistance is provided so that the maximum current of the LED is not exceeded. If there is any likelihood that the polarity of the bias voltage may be reversed, a protection diode is provided so that reverse voltage breakdown of the LED does not occur.

High electroluminescent efficiency has been obtained from GaAsP on GaP LED optimised for maximum quantum efficiency at 670 nm wavelength. A typical spectral response is shown in figure 2.5, having a spectral width of approximately 40 nm. The LED emits a narrow beam of light with an angular width of about 22 ° (see figure 2.6). The radiant intensity (power per unit solid angle) is 500 μW/sr with a forward current of 10 mA. The maximum steady

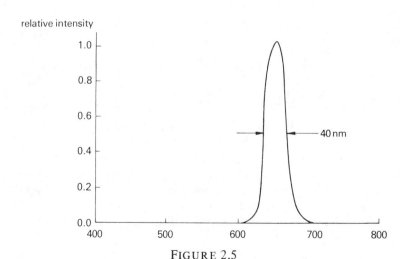

FIGURE 2.5
Spectral emission of a GaAsP on GaP LED

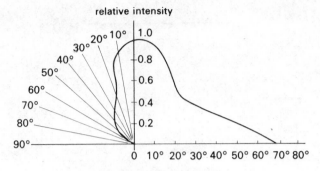

FIGURE 2.6

Radiation pattern of a GaAsP on GaP LED

forward current which this LED can withstand is 30 mA. However, if the current supply is pulsed the maximum peak current can be higher, as long as the average current does not exceed maximum levels. For instance, if the current pulses are of 10 μs duration with a repetition rate of 30 μs the peak current can be as high as 60 mA. The average current in this case is 18 mA, lower than the maximum direct current. The output intensity of the LED with a direct current of 30 mA is three times its rated value at 10 mA, while with 60 mA current pulses its output intensity is nine times the rated value. If the LED is operated with a forward current of 10 mA, its forward resistance is 1.9 V. With a supply voltage of 10 V, the series resistance needed in the bias circuit is

$$R_b = \frac{(V_{battery} - V_{diode})}{I_f}$$

$$= \frac{(10.0 - 1.9)}{10 \text{ mA}} = 810 \ \Omega$$

2.3   LED Modulator

Over a wide range of currents the light output intensity of an LED is proportional to the forward current through the device, making it highly suitable for linear modulation of light intensity. Such an LED may be used in an optical communication link or as an optically coupled isolator. In designing a modulation circuit it must be remembered that the forward voltage of the LED changes only slightly with forward current, and therefore it must be driven from a current source. A suitable circuit using a transistor as the current source is shown in figure 2.7.

The modulator input is a.c. coupled through capacitor C, which determines the low frequency cut-off of the modulator. The variable resistor R_2 is provided

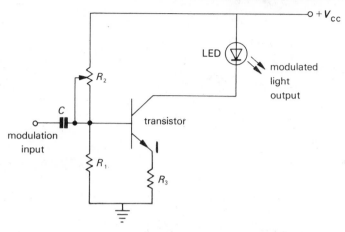

FIGURE 2.7
LED modulation circuit

so that with zero input voltage the LED forward current may be set at the
appropriate level (10 to 20 mA). The bias point should be chosen so as to give
maximum current swing. The resistor R_3 prevents the maximum forward current
of the LED being exceeded when the transistor is saturated.

In order to obtain maximum bandwidth from the LED modulator, the
transistor should be chosen so as to have response times compatible with that of
the LED. Typically, the LED response time ranges from 50 ns to 200 ns, giving
5 MHz to 20 MHz bandwidth potential.

2.4 LED Optically Coupled Isolators

An important application of the LED is in electrically decoupling different
portions of a circuit. Two circuits can be completely electrically isolated,
avoiding any feedback between output and input. This is accomplished through
an integrated circuit chip with an LED and photodiode in proximity to one
another, as shown in figure 2.8. In the optically coupled isolator a linear
relationship between input and output current is achieved over a range of input
currents from 1 mA to 30 mA. The frequency response depends on the LED and
photodiode response times, and megahertz bandwidths are easily achieved (see
figure 2.9).

The d.c. voltage between the LED cathode terminal and the output transistor
ground terminal may be as high as 3000 V, providing very effective electrical
isolation. The optically coupled isolator may be used in situations where
information must be transmitted between switching circuits electrically isolated
from one another, such as digital data transmission systems. Traditionally, this
isolation has been provided by relays, isolation transformers, line drivers and

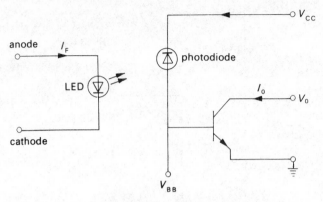

FIGURE 2.8
Circuit diagram of an optically coupled isolator

receivers. The optically coupled isolator has the advantages of very wide
bandwidth and high common mode voltage isolation. Other applications are in
sensing circuits, patient monitoring equipment, power supply feedback, high
voltage current monitoring, adaptive control systems, and audio and video
amplifiers.

2.5 LED Lamps and Displays

A wide variety of solid state lamps is available for use as indicators for appliances,
automobile instrument panels, and the like. These emit green, yellow or red
light, depending on the material and doping of the LED. The green LEDs are

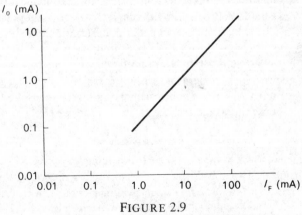

FIGURE 2.9
Output current against input current

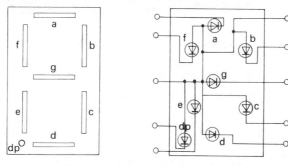

FIGURE 2.10
Typical seven-segment display

made from GaP, while the yellow and red LEDs are GaAsP on GaP. Two different red wavelengths are available at 635 nm and 655 nm. The 635 nm red LED is preferable since the human eye is more responsive at this wavelength than at 655 nm. The solid state lamps operate at a forward current of 10 to 50 mA with electrical power dissipation around 20 mW. The emitted light has a radiant intensity around 2 μW per steradian in the green, 4 μW/sr in the yellow, and 20 μW/sr in the red (635 nm). The light is emitted over a cone of angles of $\pm45°$, representing a solid angle of about 2 sr. Consequently, the total power emitted by a typical red solid state lamp at 635 nm wavelength is about 40 μW or 6 mlm. The terms used here are explained in appendix A.

LED displays are commonly used in pocket calculators, instrument readout, point of sale terminals, clocks and appliances. The seven-segment displays consist of an integrated circuit containing eight LEDs which are optically magnified to form seven individual segments plus a decimal point (see figure 2.10). They are available in red, yellow and green. Their electrical characteristics and radiant intensity are similar to those of the solid state lamps mentioned above. Numeric indicators, which consist of a monolithic GaAsP chip of eight seven-segment displays, are used in pocket calculators. These are compatible with the digital output of the calculator chip. Alphanumeric indicators using dot matrix arrays are also available, and are designed to be compatible with BCD logic inputs.

3 Solid State Photodetectors – The Photoconductor

3.1 Introduction

In this chapter we shall discuss the main detector of infrared illumination, the photoconductor. This device is of importance because it is compact, operates on a low bias voltage, and responds over a wide range of wavelengths depending on the material chosen. The photoconductor has the potential of high current response since its gain factor may be larger than unity, provided material parameters are chosen carefully. Besides infrared detection, the photoconductor finds application as a detector of intense light, in the laser power meter, and in radiometric and photometric calibration of light sources such as the LED.

3.2 The Photoconductor

At room temperature an insulator has an empty conduction band and filled valence band. If it is irradiated by photons of sufficient energy ($hf > E_g$), transitions from the valence band to the conduction band will occur, and the insulator will conduct. This process is called photoconduction.

Since a semiconductor at room temperature has a partially filled conduction band, its conductivity is increased when it is irradiated by photons of sufficient energy. A photoconductor is a block of semiconductor with metallic contacts. If a voltage is applied between the contacts current will flow, and incident light will cause an increase in this current. In the absence of any light the current which flows is referred to as the dark current, while the current which results from incident light may be referred to as the photon induced current or photocurrent.

Two types of photoconductor are possible (see figure 3.1). In the intrinsic semiconductor a photon may release an electron–hole pair, and both can contribute to the current. In the extrinsic semiconductor, only one type of carrier is available for conduction, electrons for n-type material and holes for p-type. The spectral response of extrinsic semiconductor may be modified by doping since donor or acceptor levels fall between the conduction and valence bands. The same technique is used in modifying the emission of the LED, where various colours may be produced by choice of material and doping. The energy gap and critical wavelength for several different materials is given in table 3.1 below.

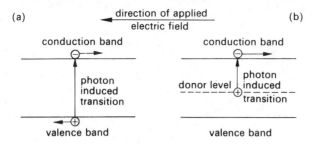

FIGURE 3.1

Generation of an electron–hole pair in: (a) intrinsic and (b) extrinsic semiconductor

3.3   The Photoconduction Process

Under dark conditions the conductivity of intrinsic semiconductor is

$$\sigma = e(n\mu_n + p\mu_p) \tag{3.1}$$

where n is the average electron concentration, p is the average hole concentration, and μ_n and μ_p are the respective electron and hole mobilities. The conductivity of extrinsic semiconductor may be obtained by setting n or p equal to zero, depending on whether the material is p-type or n-type. Suppose a sample of intrinsic semiconductor is irradiated by photons of sufficient energy to cause carrier transitions from valence to conduction band, producing N electron–hole pairs per second. If the average lifetime of electrons and holes is τ_n and τ_p, then

TABLE 3.1

Long wavelength limit of various materials

Material	E_g (eV)	λ_c (nm) ($\lambda < \lambda_c$)	
Indium antimonide (InSb)	0.16	7750	
Lead sulphide (PbS)	0.41	3020	
Germanium (Ge)	0.66	1880	Infrared
Silicon (Si)	1.12	1110	
Gallium arsenide (GaAs)	1.43	870	
Cadmium selenide (CdSe)	1.70	730	Red
Gallium phosphide (GaP)	2.24	550	Green
Cadmium sulphide (CdS)	2.42	510	

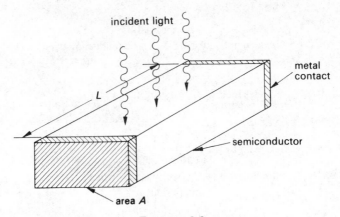

FIGURE 3.2

An intrinsic semiconductor photoconductor of volume $V = AL$

the average concentration of conduction electrons is increased by $\Delta n = N\tau_n/AL$, while the average concentration of holes is increased by $\Delta p = N\tau_p/AL$, where AL is the volume of the material (as shown in figure 3.2).

The conductance of a sample of semiconductor of length L and cross sectional area A is $G = \sigma A/L$. The increase in conductance due to the incident light is

$$\Delta G = \frac{e(\Delta n\mu_n + \Delta p\mu_p)A}{L} = \frac{eN}{L^2}(\mu_n\tau_n + \mu_p\tau_p) \tag{3.2}$$

If voltage V is applied between contacts, the increase in current above the dark current is

$$I = V\Delta G = \frac{eV}{L^2}N(\mu_n\tau_n + \mu_p\tau_p) \tag{3.3}$$

The gain factor of a photoconductor is the ratio of photocurrent to the amount of charge per unit time produced by the incident light

$$G = \frac{I}{eN} \qquad \text{(gain factor)} \tag{3.4}$$

for the intrinsic semiconductor discussed above the gain factor is

$$G = \frac{V}{L^2}(\mu_n\tau_n + \mu_p\tau_p) \qquad \text{(intrinsic semiconductor)} \tag{3.5}$$

A sensitive photoconductor must have a large gain factor, which occurs if the product of mobility and average carrier lifetime is large, or if the distance L between contacts is small.

A physical interpretation of the gain factor for a photoconductor may be obtained by considering each of the two terms in equation 3.5 separately. The gain factor for electrons in the photoconductor is $G_n = \mu_n \tau_n V/L^2$. The time taken for an electron to drift a distance L, the electron transit time, is $t_n = L/V_{dr}$, where V_{dr} is the electron drift velocity, given by $V_{dr} = \mu_n V/L$. Thus the electron transit time between contacts is

$$t_n = \frac{L^2}{\mu_n V} \qquad \text{(electron transit time)} \qquad (3.6)$$

and the gain factor for carrier electrons in the photoconductor is

$$G_n = \frac{\tau_n}{t_n} \qquad \text{(gain factor for electrons)} \qquad (3.7)$$

The corresponding expression for holes is

$$G_p = \frac{\tau_p}{t_p} \qquad \text{(gain factor for holes)} \qquad (3.8)$$

Since the average lifetime of electrons and holes is the same in a semi-. conductor, we set $\tau = \tau_n = \tau_p$ = average carrier lifetime. The gain factor for an intrinsic semiconductor is

$$G = G_n + G_p = \frac{\tau}{t_n} + \frac{\tau}{t_p} = \tau \left(\frac{1}{t_n} + \frac{1}{t_p} \right)$$

If we define the sum of $1/t_n$ and $1/t_p$ as $1/t_{dr}$ where t_{dr} is the effective time taken for carriers to drift between the contacts of the photoconductor

$$\frac{1}{t_{dr}} = \frac{1}{t_n} + \frac{1}{t_p}$$

then the gain factor becomes

$$G = \frac{\tau}{t_{dr}} \qquad \text{(gain factor of a photoconductor)} \qquad (3.9)$$

This expression for the gain factor of a photoconductor is valid for both intrinsic and extrinsic semiconductors, provided the appropriate value for t_{dr} is used.

It is possible for a photoconductor to have a gain factor larger than unity. This occurs if the average carrier lifetime is larger than the time taken for charges to drift between contacts. In a photoconductor which is irradiated by light, electron—hole pairs are spontaneously created. As long as a voltage is applied across the material, these charges drift towards the contacts at either end of the piece of semiconductor. Since the material must maintain charge neutrality,*

* A semiconductor can have net charge for a transient period τ_d equal to the dielectric relaxation time for the material. For a semiconductor with dielectric constant ϵ and conductivity σ the dielectric relaxation time is $\tau_d = \epsilon/\sigma$. For example, germanium has $\epsilon = 16\epsilon_0$, $\sigma = 100$ $(\Omega \text{ m})^{-1}$ giving $\tau_d = 1.4 \times 10^{-12}$ s. Since this is much shorter than τ or t_{dr}, it is correct to say that charge neutrality is maintained in the material.

when an electron passes into one of the contacts a second electron must come from the opposite contact into the material. Thus carriers are not lost into the contacts through carrier drift. By making the distance L between contacts small enough, it is possible for the carrier transit time to be smaller than the average carrier lifetime, establishing a gain factor larger than unity.

3.4   Quantum Efficiency

In an intrinsic photoconductor an incident photon may release one electron–hole pair, or it may not, depending on its wavelength and other factors. The wavelength of the incident photon must be sufficiently short so that it gives up enough energy, greater than the energy gap between valence and conduction band. On the other hand, if it has too much energy it may pass straight through the semiconductor or it may be absorbed too near the surface (surface charges have a shorter average lifetime than those inside). The probability that an incident photon will produce an electron–hole pair in a photoconductor is expressed by its quantum efficiency, η. The quantum efficiency varies with wavelength (see figure 3.3) and is always less than unity, since we do not expect a single photon to release two electron–hole pairs.

If a photoconductor collects light over a total surface area A_c, and if the number of photons per unit area per unit time which falls on the collection

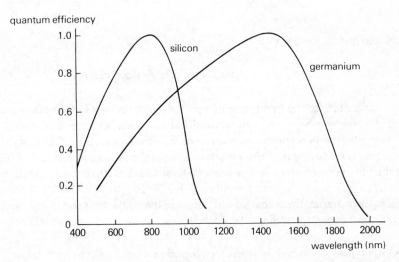

FIGURE 3.3

The variation of the quantum efficiency of silicon and germanium with wavelength

surface is N_0, then the d.c. photocurrent produced is

$$I = \eta e N_0 A_c G \qquad (3.10)$$

where G is the gain factor for the photoconductor, as in equation 3.5. In equation 3.10 it is assumed that light of a single wavelength is being absorbed, and η is the specified quantum efficiency at that wavelength. If a photoconductor is illuminated by white light it is necessary to determine the average quantum efficiency by integrating η times the emission spectrum of the white light source over all wavelengths. The average quantum efficiency will always be less than the maximum value for monochromatic light.

3.5 Photoconductor Responsivity

The d.c. photocurrent produced by $N_0 A_c$ photons per second in a photoconductor with quantum efficiency η and gain factor G is $I = \eta e N_0 A_c G$, as in equation 3.10. It is more convenient, however, to express the incident light in terms of its power density or total power, rather than number of photons per second. Since each photon has energy equal to hf J, the power represented by $N_0 A_c$ photons per second is

$$P_{in} = N_0 A_c hf \text{ J/s} \qquad \text{(watts)} \qquad (3.11)$$

The d.c. photocurrent produced by an illumination with total power P_{in} is

$$I = \eta e \frac{P_{in}}{hf} G$$

Since the energy of a photon is usually expressed in electron volts, hf/e, the d.c. photocurrent produced by P_{in} is

$$I = \frac{\eta G P_{in}}{E_0}, \quad E_0 = \frac{hf}{e} \qquad (3.12)$$

The responsivity of a photoconductor is defined as the ratio of photocurrent to incident illumination power.

$$R_\phi = \frac{I}{P_{in}} = \frac{\eta G}{E_0} \qquad \text{(photoconductor responsivity)} \qquad (3.13)$$

where the responsivity has units of amperes per watt. For example, a photoconductor with a quantum efficiency of $\eta = 1$ at its peak wavelength, $\lambda = 800$ nm, and a gain factor $G = 1$ has a responsivity of $R_\phi = 0.643$ A/W, since the energy of a photon at that wavelength is 1.555 eV.

Although potentially much higher responsivities are possible in the photoconductor if the gain factor is increased, in practice responsivities of less than 1 A/W are the usual case. This may be due to the fact that the distance L between contacts which produces a high gain factor is so small that the collection area of the photoconductor is too small to be practicable. The gain factor may also be increased by making the average lifetime of carriers as long as possible, but this causes a rather slow response for the photoconductor, as we shall see in the next section.

3.6   Frequency Response of a Photoconductor

If a single photon is absorbed in an intrinsic photoconductor releasing an electron–hole pair, a pulse of current is produced which lasts until the carriers are trapped. The amplitude of current which results is calculated as follows. If the volume of the photoconductor is AL, where A is the cross sectional area of a contact and L the distance between contacts as in figure 3.3, the charge density which results from a single electron is e/AL. The current is the charge density times the drift velocity of the electron, V_n, times the area A, $I_e = eV_n/L$. But the time for an electron to drift between contacts is $t_n = L/V_n$. Thus the current due to a single electron is e/t_n, while the current due to a hole is e/t_p, where t_p is the time taken for holes to drift between contacts. If the carriers last a time T before being recaptured, the amplitude of the current pulse produced by one electron–hole pair is $e(1/t_n + 1/t_p)$ (see figure 3.4).

This result applies to a single event, but we would like to know what happens on average. There is an average lifetime of free charges in a semiconductor, τ. We need to consider the statistics of the carrier lifetime, T.

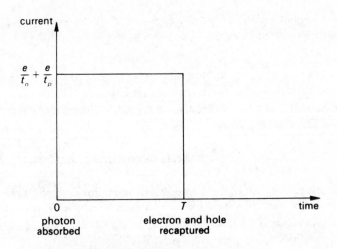

FIGURE 3.4

Current pulse produced by the absorption of one photon, assuming the electron and hole are recaptured at time T

We could study the statistics of charge recombination by observing individual charge release and recapture events. Suppose we observe 100 such events, recording the time taken for an electron–hole pair to be recaptured. We could compute the distribution of carrier lifetimes and the average carrier lifetime from these observations. Fortunately, this experiment has already been done. The statistics of charge recombination are expressed by the probability that an electron–hole pair released at time $t = 0$ will be recaptured in the time interval T to $T + dT$, which is

$$p(T)dT = \frac{1}{\tau} \exp(-T/\tau) \, dT \qquad \text{(probability that a free charge will be recaptured in time interval } T \text{ to } T + dT) \qquad (3.14)$$

In this expression τ is the average carrier lifetime.

The probability that a charge released at time $t = 0$ will have recombined by time t_1 is

$$P(t_1) = \int_0^{t_1} p(T) \, dT = \frac{1}{\tau} \int_0^{t_1} \exp(-T/\tau) \, dT = 1 - \exp(-t_1/\tau) \qquad (3.15)$$

For example, if the average lifetime of free charges is $\tau = 1 \ \mu s$, the probability that a charge will have been recaptured in $1 \ \mu s$ is $1 - \exp(-1) = 0.63$, and in $5 \ \mu s$ the probability of recapture is $1 - \exp(-5) = 0.9933$. A probability of unity means absolute certainty of recapture (see figure 3.5).

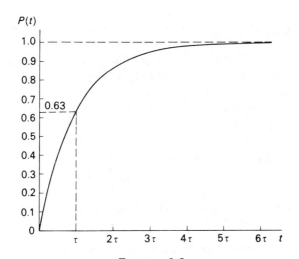

FIGURE 3.5

Probability that a charge released at $t = 0$ will have been recaptured in time t

The method of calculating the frequency response of a photoconductor employed here is to calculate the average impulse response $H(s)$. Then the steady state response to a sinusoidal variation in light intensity of frequency ω is $H(s = j\omega)$ times the total incident light power at frequency ω.

The Laplace transform of the current pulse which results from absorption of a single photon is

$$\int_0^T \left(\frac{e}{t_n} + \frac{e}{t_p} \right) \exp(-st) \, dt = \left(\frac{e}{t_n} + \frac{e}{t_p} \right) \frac{1 - \exp(-sT)}{s} \qquad \text{(Laplace transform of current pulse)}$$

The average pulse response is

$$H(S) = \int_0^\infty p(T) \left(\frac{e}{t_p} + \frac{e}{t_n} \right) \frac{1 - \exp(-sT)}{s} \, dT$$

$$= \left(\frac{e}{t_n} + \frac{e}{t_p} \right) \frac{1}{s} \int_0^\infty p(T)[1 - \exp(-sT)] \, dT$$

$$= e \left(\frac{\tau}{t_n} + \frac{\tau}{t_p} \right) \frac{1}{1 + s\tau} \qquad (3.16)$$

Recalling that the gain factor is $G = \tau/t_n + \tau/t_p$ and multiplying by the quantum efficiency η, and the incident power at frequency ω, the photocurrent at frequency $s = j\omega$ is

$$I(\omega) = \frac{\eta P_{\text{inc}}(\omega)}{hf} H(j\omega) = \eta \frac{P_{\text{inc}}(\omega)}{E_0} \frac{G}{1 + j\omega\tau} \qquad (3.17)$$

where E_0 is the energy of incident photons expressed in electron volts $(E_0 = hf/e)$.

This result shows that the frequency response of a photoconductor is limited by the average carrier lifetime, as one might expect from intuition.

If a photoconductor is used as a detector in an optical communication system where the modulation signal has a spectrum centered at frequency ω_m, then assuming $\omega_m \tau \gg 1$, the response at this frequency is

$$I(\omega_m) = \eta \frac{P_{\text{in}}(\omega_m)}{E_0} \left(\frac{\tau}{t_n} + \frac{\tau}{t_p} \right) \frac{1}{j\omega_m \tau}$$

$$= \eta \frac{P_{\text{in}}(\omega_m)}{E_0} \left(\frac{1}{j\omega_m t_n} + \frac{1}{j\omega_m t_p} \right)$$

Thus at high frequencies the photoconductor response depends only on the time taken for carriers to drift between contacts, the dependence on average carrier lifetime disappears altogether.

3.7   Equivalent Circuit of a Photoconductor

If a photoconductor is irradiated by light which is modulated at frequency ω_m, the incident power is

$$P_{in} = P_0 + P_1 \exp(j\omega_m t)$$

The resulting photocurrent is

$$I = \eta \frac{P_0}{E_0} G + \eta \frac{P_1}{E_0} \frac{G}{1 + j\omega_m \tau} \exp(j\omega_m t)$$

$$= I_0 + \frac{I_1}{1 + j\omega_m \tau} \exp(j\omega_m t)$$

where $I_0 = \eta P_0 G/E_0$ is the d.c. photocurrent and $I_1/(1 + j\omega_m \tau)$ is the photocurrent at frequency ω_m.

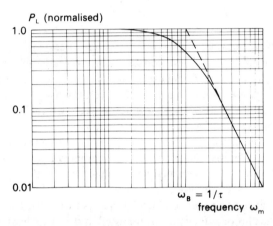

FIGURE 3.6

Power available versus modulation frequency of incident illumination

The equivalent circuit of the photoconductor at frequency ω_m is a current source shunted by a resistance R (see figure 3.7). The resistance R is just the bulk resistance of the semiconductor, $R = L/\sigma A$, where L is the distance between contacts, A the cross sectional area of a contact, and σ the conductivity of the material.

The maximum power obtainable from a photoconductor may be calculated by computing the power delivered to a matched load resistance $R_L = R$. The current through a matched load resistance is

$$I_L = \tfrac{1}{2} \frac{I_1}{1 + j\omega_m \tau}$$

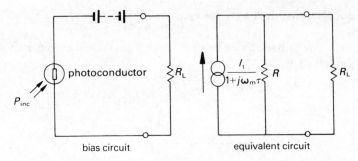

bias circuit   equivalent circuit

FIGURE 3.7

Photoconductor bias circuit and equivalent circuit

The power delivered to the load resistance is

$$P_{av} = |I_L|^2 R = \tfrac{1}{4} \frac{I_1^2 R}{1 + (\omega_m \tau)^2} \tag{3.18}$$

In this expression $I_1 = \eta P_1 G / E_0$ is the rms amplitude of current.

We see in figure 3.8 that the photoconductor behaves rather like a current source shunted by a resistance R and capacitance $C = \tau/R$, where τ is the average lifetime of carriers in the photoconductor. However, C must not be confused with the shunt capacitance of the photoconductor, C_{sh}, which for dimensions A and L and semiconductor dielectric constant ϵ would be $C_{sh} = \epsilon A/L$. Since the shunt resistance of the device is $R = L/\sigma A$, $C_{sh} = \epsilon/\sigma R$. Using the dielectric relaxation time $\tau_d = \epsilon/\sigma$, the shunt capacitance is given by $C_{sh} = \tau_d/R$. Since the dielectric relaxation time of a semiconductor is many orders of magnitude smaller than the average carrier lifetime, it may be concluded that $C_{sh} \ll C = \tau/R$ (the shunt capacitance of a photoconductor is negligibly small). This result will not be true for the photodiode, which is discussed in the next chapter.

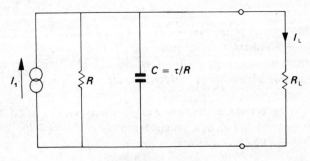

FIGURE 3.8

Another photoconductor equivalent circuit

4 Solid State Photodetectors– The Photodiode and Phototransistor

4.1 Introduction

This chapter discusses the two types of photodetector found most commonly in optoelectronic systems, the photodiode and phototransistor. Both devices are compact light detectors requiring no more than a few tens of volts for bias, and if made of silicon are responsive to light in·the visible and near infrared·wavelengths. Improved device technology has increased the sensitivity and bandwidth of the photodiode to the point where its performance is comparable with that of the photomultiplier tube, which is sensitive and fast enough to detect individual photons. The avalanche photodiode is particularly similar in performance to the photomultiplier tube. The photodiode finds application in the optically coupled isolator, optical data links, and in optical communications, where a fast response time is required for the very wide bandwidth expected of an optical communication system. The phototransistor, although somewhat more limited in bandwidth than the photodiode, finds numerous applications as a high-current response photodetector.

4.2 The Depletion Region Photodiode

A photodiode consists of a reverse biased $p-n$ junction as shown in figure 4.1. Photons absorbed in the depletion region of the junction release electron–hole pairs, which are swept across the junction, resulting in a steady photocurrent. As in the case of the photoconductor, the incident photons must have sufficient energy to bridge the gap between valence and conduction band for the material. In the PIN photodiode of figure 4.1, the thickness of the depletion region is controlled by using a layer of intrinsic semiconductor between the p-type and n-type layers. The p-layer is very thin so that a negligible number of photons are absorbed in this region. The depletion region is thick enough to capture most of the incident light, but not so thick that the transit time for carriers across this region is too large, so as to reduce the bandwidth of the device. The gain factor of a photodiode is unity since electrons and holes are immediately recaptured upon drifting out of the depletion region.

As light travels through a semiconductor it is absorbed, either contributing energy to mechanical vibrations or releasing carriers for conduction. If the light

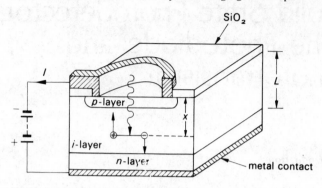

FIGURE 4.1
Reverse biased silicon PIN photodiode

enters the material with intensity I_0, its intensity after travelling a distance x is $I_0 \exp(-\alpha x)$, where α is the attenuation constant for the material. The attenuation constants for silicon and germanium are plotted against wavelength in figure 4.2. At a wavelength of 800 nm the attenuation constant of silicon is $\alpha = 10^3$ cm^{-1}. The effective depth of penetration of light with that wavelength in silicon is $1/\alpha = 10^{-3}$ cm = 10 μm.

The formula for the absorption of light in a semiconductor may be interpreted in a different way. If we consider a single photon entering a piece of semiconductor it will travel unaffected into the material until it is absorbed, since

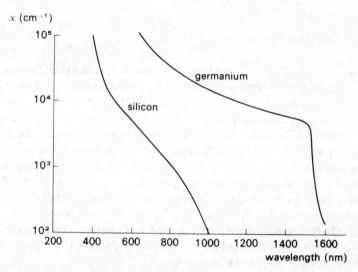

FIGURE 4.2
Absorption constant of silicon and germanium at 300 K

there is no gradual attenuation of its energy. The probability that a photon will have been absorbed before reaching a depth x into the material is $1 - \exp(-\alpha x)$. The probability density for photon absorption is

$$p(x) = \frac{d}{dx} [1 - \exp(-\alpha x)] = \alpha \exp(-\alpha x) \tag{4.1}$$

The interpretation of $p(x)$ is that the probability that a photon will be absorbed between x and $x + dx$ in the material is $p(x)\,dx$. (Recall a similar definition for carrier lifetime in section 3.6.)

In order to calculate the photocurrent produced in a reverse biased photo-diode it is necessary to consider the attenuation constant of the depletion region of the device. In the case of the PIN photodiode of figure 4.1 this is the attenuation constant of silicon. Obviously it is more likely that photons are absorbed in the depletion region and less likely that they survive to reach the n-layer below (see figure 4.3). If a photon enters the depletion region and is absorbed at a distance x into it, releasing an electron–hole pair, the result is two pulses of current, one for the electron and one for the hole. If the depletion region is of width L, the hole must travel a distance x to reach the p-layer where it is recaptured, while the electron must travel a distance $L - x$ to be recaptured in the n-layer.

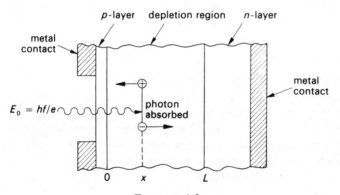

FIGURE 4.3
The mechanism of the absorption of a photon in the depletion region of a silicon PIN photodiode

Consider first the pulse of current which results from the hole. If the drift velocity of the hole is v_p, the time for a hole to drift across the depletion region is $t_p = L/v_p$. If the hole travels a distance x to be captured in the p-layer the time taken is $t_p x/L = x/v_p$. This produces a pulse of current of amplitude e/t_p and duration x/v_p as shown in figure 4.4. Since we are interested in the average

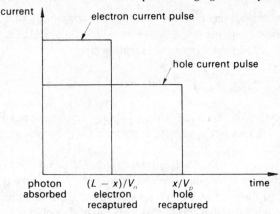

FIGURE 4.4

Current pulses produced by an electron–hole pair at distance x into the depletion region. Here $t_p = L/v_p$ and $t_n = L/v_n$, v_p and v_n are the hole and electron drift velocities

current in the photodiode, the net charge of this current pulse is

$$\frac{e}{t_p} \cdot \frac{x}{v_p} = \frac{ex}{L} \quad \text{(coulombs)} \tag{4.2}$$

In the same way, the net charge produced by the electron, which travels a distance $L - x$ to be captured in the n-layer is

$$\frac{e(L - x)}{L} \quad \text{(coulombs)} \tag{4.3}$$

The total net charge from the electron–hole pair released at distance x into the depletion region of the photodiode is the sum of equations (4.2) and (4.3), which is $ex/L + e(L - x)/L = e$ coulombs. This result is not surprising. Consider the extreme case of a photon absorbed at $x = 0$ in the depletion region. Of the resulting electron–hole pair the hole is immediately recaptured, producing no current, while the electron travels a distance L before being recaptured, producing a current pulse of amplitude e/t_n and duration L/v_n, and net charge of e coulombs.

If there are P_{in}/hf photons per second incident on the depletion region of the PIN photodiode, the average photocurrent produced is

$$I = \int_0^L \quad \eta \quad e \quad \frac{P_{in}}{hf} \quad \alpha \exp(-\alpha x) \; dx \tag{4.4}$$

$$\underbrace{\qquad}_{\substack{\text{quantum} \\ \text{efficiency}}} \underbrace{\qquad}_{\substack{\text{total} \\ \text{net} \\ \text{charge}}} \underbrace{\qquad}_{\substack{\text{number} \\ \text{of} \\ \text{photons} \\ \text{per second}}} \underbrace{\qquad}_{\substack{\text{probability} \\ \text{density} \\ \text{for absorption}}}$$

Since none of the factors in the integrand depend on x, except for $p(x)$, the average photocurrent is

$$I = \eta[1 - \exp(-\alpha L)] \frac{P_{in}}{E_0} \qquad \left(E_0 = \frac{hf}{e}\right) \qquad (4.5)$$

This formula is correct for a photodiode with a depletion region of thickness L. It has a very simple interpretation. As we have seen, an electron–hole pair created in the depletion region produces two pulses of current whose total net charge is e coulombs. The probability that a photon of energy $E_0 = hf/e$ electron volts will have been absorbed before reaching the n-layer is $1 - \exp(-\alpha L)$. If the total incident power is P_{in} watts then the average photocurrent is given by equation 4.5, since the number of photons incident is P_{in}/hf per second.

This result differs from that of the photoconductor in two ways. First the gain factor of a photodiode is unity since all carriers are captured immediately upon leaving the depletion region. Secondly, in the photoconductor all photons are absorbed in the material, while in the photodiode only a portion $1 - \exp(-\alpha L)$ are absorbed in the depletion region. We could say that the effective quantum efficiency of a photodiode is

$$\eta' = \eta[1 - \exp(-\alpha L)] \qquad \text{(effective quantum efficiency of a photodiode)} \qquad (4.6)$$

In analysing the behaviour of the PIN photodiode we have ignored the possibility that a photon will be absorbed in the p-layer or the n-layer. If this happens charges may be released which will diffuse toward the depletion region of the photodiode, contributing a small current. The p-layer is very thin and the probability of the absorption of a photon in this region is small. However, high energy photons may be absorbed since for sufficiently short wavelengths the absorption constant is very large. No matter how thin the p-layer is, there must be a wavelength below which it is thicker than $1/\alpha$. It is found in practice that although photons with wavelengths between 400 nm and 1100 nm are not absorbed in the p-layer of the photodiode, photons in the ultraviolet spectrum below 400 nm may be, and diffusion current may be an important factor in this part of the spectrum. A similar argument shows that photons of sufficiently long wavelength may survive to be absorbed in the n-layer, so that diffusion current may be an important factor in the infrared portion of the spectrum.

4.3 Current Responsivity of a PIN Photodiode

The response of a photodiode may be expressed in two different ways. The ratio of output current to input power of illumination is the current responsivity

$$R_\phi = \frac{I_p}{P_{inc}} = \frac{\eta'}{E_0} \qquad \text{(current responsivity)} \qquad (4.7)$$

For example, a silicon PIN photodiode which has a peak spectral response at a wavelength of 770 nm has a current responsivity of 0.5 A/W. Since at this wavelength the incident photons have energy $E_0 = 1.61$ eV, the effective quantum efficiency of this photodiode is $\eta' = 0.5 \times 1.61 = 0.805$ (80.5 per cent) electrons per photon.

In some cases the intensity of the incident light (in mW/cm^2) is specified, rather than the total incident power. If the active area of the photodiode is A then the total incident power is

$$P_{\text{inc}} = P_{\text{dens}}A$$

where P_{dens} is the incident intensity. The axial incidence response is

$$R_E' = \frac{I}{P_{\text{dens}}} = \eta' \frac{A}{E_0} \qquad \text{(axial incidence response)} \qquad (4.8)$$

The PIN photodiode mentioned above has an active area of $A = 2 \times 10^{-3}$ cm^2. Thus its axial incidence response is $R_E = R_\phi A = 1$ μA per mW/cm^2.

The variation with wavelength of the current responsivity and effective quantum efficiency of a typical PIN photodiode are shown in figure 4.5. This photodiode has an efficient response at all wavelengths in the visible and near infrared portion of the optical spectrum, with cut-off near 1110 nm as expected from the bandgap of silicon (see table 2.1).

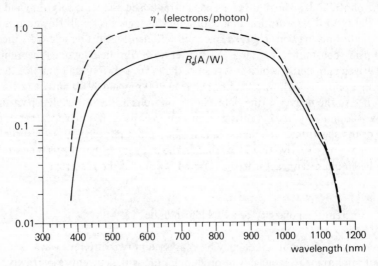

FIGURE 4.5

Equivalent circuit of a PIN photodiode. Typically $R_s = 50$ Ω, $R_p = 10^{11}$ Ω, $C_p = 5$ pF

4.4 Frequency Response of a PIN Photodiode

The frequency response of a photodiode is controlled by two main factors: the drift time of carriers across the depletion region of the device, and the junction capacitance. The reader may wish to review the discussion of carrier transit time in the photoconductor in section 3.6. In the photodiode the depletion region is thin enough that the carrier transit time across it is of the order of $t_{dr} = 10$ ps. Consequently the bandwidth of the photodiode, as determined by carrier transit time, is around 10 GHz. This figure is somewhat variable since the thickness of the depletion region of a particular photodiode varies with reverse bias voltage, a faster response being obtained at higher bias voltages. In the photoconductor the capacitance between metallic contacts is small enough to be negligible. However, the depletion region of the photodiode is thin enough that the junction capacitance C_p is important. The junction capacitance also varies with voltage since again the thickness of the depletion region increases at higher reverse bias voltages, causing a reduction in junction capacitance. Values of C_p between 1 pF and 1000 pF are typical.

The layers of p-type, intrinsic and n-type silicon of the PIN photodiode are responsible for a small series resistance, R_s. Provided the photodiode is reverse biased its junction resistance is very large, $R_p = 10^{11}$ Ω. The equivalent circuit consists of a current source in parallel with R_p and C_p, feeding the series resistance R_s. The bandwidth of the device is determined by the time constant $R_s C_p$, assuming that this exceeds the carrier transit time. Consequently, the photocurrent at frequency ω is $I_p = \eta' P_{inc}/E_0$ where P_{inc} is the incident optical power at frequency ω.

For a large bandwidth C_p must be as small as possible. Since the junction capacitance is directly proportional to the junction area and inversely proportional to the thickness of the depletion region, all high speed photodiodes have a small active area, and as thick a depletion region as possible. Since the thickness of the depletion region increases with applied reverse bias, the fastest response is obtained with high reverse bias voltage. However, if the depletion region is too thick the carrier transit time will increase to the point where the bandwidth will be limited by this effect. In practice a compromise between junction capacitance and carrier transit time gives a bandwidth of the order of 1 GHz, depending on the external circuit to which the photodiode is connected.

The power available from the photodiode may be determined using the equivalent circuit of figure 4.6. Assuming that $1/R_p$ is negligible the current delivered to the load resistance R_L is

$$I_L = \frac{I_p}{1 + j\omega(R_s + R_L)C_p} \tag{4.9}$$

The power delivered to the load is

$$P_L = |I_L|^2 R_L = \frac{I_p^2 R_L}{1 + \omega^2(R_s + R_L)^2 C_p^2} \tag{4.10}$$

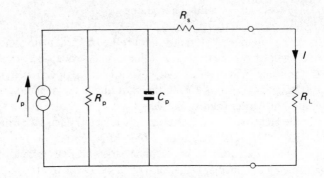

FIGURE 4.6
The current responsivity and effective quantum efficiency of a silicon PIN
photodiode

A plot of the variation of load power with frequency (figure 4.7) shows that the
3 db bandwidth of the circuit is $\omega_B = 1/(R_s + R_L)C_p$.

The highest power is extracted from the photodiode with a very large load
resistance. The power obtained when $R_L = R_p$ is

$$P = \frac{I_p^2 R_p}{4} \qquad (4.11)$$

The bandwidth with such a large load would be rather small. If $R_p = 10^{11}$ Ω,
$C_p = 5$ pF, $\omega_B = 2$ rad/s or $B = 0.32$ Hz.

We see that in order to obtain a large signal power from a photodiode it is
necessary to connect it to a very large load resistance, at the expense of signal
bandwidth. This is an example of the fact that the gain–bandwidth product of a

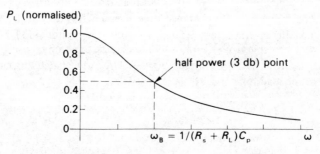

FIGURE 4.7
Frequency response of a photodiode circuit

device is constant, independent of any external factors. The gain–bandwidth product of the photodiode is inversely proportional to its junction capacitance. Methods of increasing signal power without reducing bandwidth are discussed in the next section.

4.5 The Photodiode Preamplifier

We have seen that the reverse biased PIN photodiode behaves as a current source with very large source impedance, small junction capacitance, and a small series resistance. For the purposes of signal amplification it is desirable to transform the photocurrent into a voltage with a moderate output impedance. If the rather large load resistance used with most photodiodes were connected directly to the input of an amplifier, the impedance mismatch would greatly reduce the signal. The solution to this problem is to use an operational amplifier as a current amplifier, as in figure 4.8.

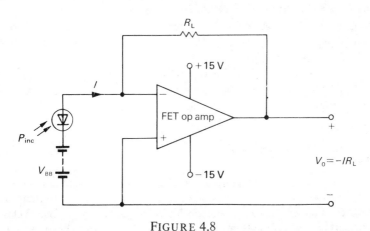

FIGURE 4.8

Photodiode preamplifier using an operational amplifier

Because of the high junction resistance of the reverse biased photodiode, the operational amplifier should be an FET type with very high input impedance. Consequently, the photocurrent flows through the feedback resistance R_L, which acts as the load resistance for the photodiode. Since the negative input of the operational amplifier acts as a virtual earth, the output voltage of the circuit is

$$V_0 = -I_p R_L \qquad (4.12)$$

The maximum power available from the photodiode preamplifier is obtained when the load resistance matches the output impedance of the operational amplifier

$$P_{av} = \frac{V_0{}^2}{4R_0} = \frac{I_p{}^2 R_L{}^2}{4R_0} \qquad (4.13)$$

Comparing with equation 4.10 we see that for the same load resistance and bandwidth, the output power has been increased by the factor $R_L/4R_0$. Typically $R_L = 1$ MΩ and $R_0 = 100$ Ω giving an increase in output power of 2500 times.

The photodiode FET operational amplifier preamplifier may be regarded as an impedance transformer, which transforms the high impedance current source into a low impedance voltage source. It must be remembered, however, that since the operational amplifier is acting as a high gain d.c. amplifier, subsequent stages of amplification should be a.c. coupled. In practice it is necessary to distinguish between two general uses of the photodiode.

If the photodiode is used as a detector of light intensity, as for example in a laser power meter or photometer, it is desirable that the d.c. output voltage be proportional to the photocurrent, as in the circuit of figure 4.8. Since in this case signal bandwidth is not needed, a very large feedback resistance may be used, values as high as 100 MΩ being typical in practice.

If the photodiode is used as a detector in an optical communication system or digital data link, then it is desirable to have large bandwidth as well as signal gain. In this case the size of the feedback resistance is restricted by the requirement of signal bandwidth. Further signal amplification is provided by a second capacitor coupled stage, which acts as an a.c. amplifier.

4.6   Photovoltaic Operation of the Photodiode

The advantage of operating a photodiode with reverse bias, or in the photoconductive mode as it is sometimes called, is that the junction capacitance is reduced and therefore its gain—bandwidth product is increased. Nevertheless, in many situations the external lead capacitances are larger than the diode junction capacitance. In these situations the additional supply needed to bias the photodiode cannot be justified, and it is operated without benefit of any external bias voltage, as in figure 4.9. This is usually referred to as photovoltaic operation of the photodiode.

In the photovoltaic mode the photocurrent flowing through the load resistance causes the photodiode junction to be forward biased, reducing the load current through forward junction current. Photovoltaic operation of the photodiode may be analysed by means of the circuit of figure 4.10, which shows a current source representing the photocurrent in parallel with the idealised diode, both in series with the diode series resistance. If the junction voltage is V_d, the load current is

$$I = I_p - I_s[\exp(eV_d/kT) - 1] \qquad (4.14)$$

$$I = I_p - I_s [\exp(eV_d/kT) - 1]$$

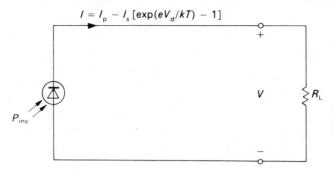

FIGURE 4.9

Photovoltaic operation of a photodiode

where I_s is the reverse bias saturation current. The diode junction voltage is

$$V_d = V + IR_s \qquad (4.15)$$

These two equations may be solved for the relationship between external voltage and current, which is

$$V = -IR_s + V_0 \ln [1 + (I_p - I)/I_s] \qquad (4.16)$$

where $V_0 = kT/e = 25.8$ mV at $T = 300$ K.

A plot of the current–voltage characteristics of a photodiode operated in the photovoltaic mode, as given in equation 4.16, with $R_s = 50 \ \Omega$ and $I_s = 5$ nA, is shown in figure 4.11. Since the current is a few tens of microamperes, the voltage drop across R_s is negligible compared with the external voltage of the photodiode. Also shown in the figure is a load line whose slope is the inverse of the load resistance. For a given photocurrent, the operating point is the intersection of the load line and the current–voltage characteristic.

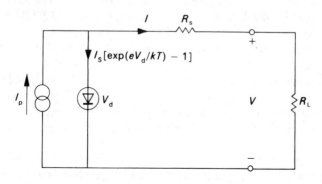

FIGURE 4.10

Circuit representation of a photodiode operated in the photovoltaic mode

external current (μA)

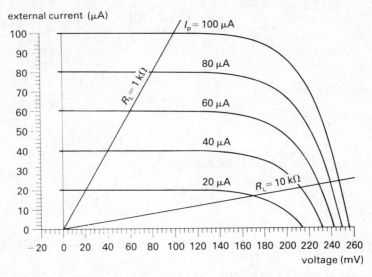

FIGURE 4.11
Current—voltage characteristics of a photodiode operated in the photovoltaic
mode, $R_s = 50\ \Omega, I_s = 5$ nA

The dynamic resistance of the photodiode at the operating point is the inverse of the rate of change of diode forward current with diode voltage. Thus

$$R_p = \left(\frac{V_0}{I_s}\right) \exp\left(-\frac{V_d}{V_0}\right) \tag{4.17}$$

In general the dynamic resistance is smaller in the photovoltaic mode than in the photoconductive mode, as may be seen in figure 4.11. For example, with $V_0 = 25.8$ mV, $I_s = 5$ nA, and $V_d = 25.8$ mV, $R_p = 5.16/e = 1.9$ MΩ, as compared with 10^{11} Ω in the photoconductive mode.

The small signal equivalent circuit of the photodiode operated in the photo-photovoltaic mode is the same as that of figure 4.6, except that the resistance R_p is lower. The approximation of a linear equivalent circuit is only valid for small changes of voltage and current around the operating point, since in the photovoltaic mode the external current varies approximately as the logarithm of photocurrent. This may be seen by substituting $V = R_L I$ in equation 4.16 and solving for I_p in terms of I.

Linear response of the photodiode operated in the photovoltaic mode is restricted to low values of photocurrent, depending on how large is the load resistance. A small load resistance gives a much wider range of linear response. A general rule for linear response is $I_{max} < V_0/R_L$.

The primary advantage of photovoltaic operation of a photodiode is that the dark current is zero. Thus very low levels of light may be detected. Disadvantages are that the junction capacitance is increased, restricting bandwidth, and that linear operation is possible only over a restricted range of photocurrent.

4.7 The Avalanche Photodiode

Although the PIN photodiode is the most commonly available photodiode configuration, there are many other types for special applications. The Schottky barrier silicon photodiode consists of a gold film metal to semiconductor junction, and is capable of a very large active area and of response to short wavelengths below 300 nm in the ultraviolet portion of the spectrum. The electrical characteristics of the Schottky barrier photodiode are similar to those of the PIN photodiode already discussed, and the equivalent circuit is the same as the one in figure 4.6. The avalanche silicon photodiode gives current multiplication through reverse voltage breakdown, similar to the current multiplication of the traditional photomultiplier tube. When the device is reverse biased very near to the reverse breakdown voltage, an incident photon releases an electron which releases additional electrons through impact ionisation. In the configuration shown in figure 4.12, holes created in the high electric field region where the breakdown occurs (in the depletion region of the $p-n$ junction) drift to the

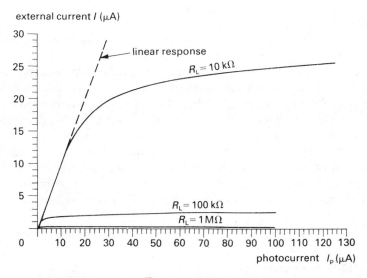

FIGURE 4.12
The variation of external current with photocurrent for different values of load resistance, for photovoltaic operation of the photodiode

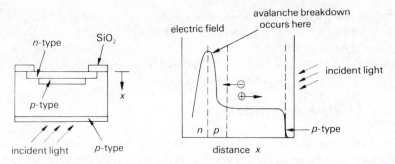

FIGURE 4.13

Configuration of a silicon avalanche photodiode

heavily doped p^+ layer, constituting the multiplied photocurrent. The equivalent circuit of an avalanche photodiode consists of a current source with multiplication factor M, junction capacitance C_p and series resistance R_s, as shown in figure 4.13. The power delivered to a load resistance R_L is

$$P_L = \frac{M^2 |I|^2 R_L}{1 + \omega^2 (R_s + R_L)^2 C_p^2} \qquad (4.18)$$

Comparison with equation 4.10 shows that the power available from an avalanche photodiode is M^2 times that available from a PIN photodiode.

Typical values for an avalanche photodiode are $M = 100$ with a responsivity of 63 A/W, quantum efficiency of 85 per cent, junction capacitance of 4 pF and $R_s = 50 \ \Omega$, with reverse breakdown voltage in the range 230 to 330 V. It is interesting to compare the avalanche photodiode with a standard photomultiplier tube such as the S–20, Na_2K Sb–Cs type which has peak spectral response at 400 nm. The peak flux responsivity of the S–20 photomultiplier is 600 A/W with a bias voltage around 1500 V. This response falls to about 230 A/W at 600 nm and 31 A/W at 800 nm. Since the peak response of the silicon avalanche photodiode occurs at about 800 nm, the two devices have very comparable responses at that wavelength. The avalanche photodiode offers the advantages of compactness and low bias voltage, with electrical characteristics similar to those of the photomultiplier. However, the dark current of the avalanche photodiode is around five times higher than that of the photomultiplier; consequently the photomultiplier is a more sensitive detector of low light intensities.

4.8 The Phototransistor

It is not difficult to imagine how the phototransistor came to be invented. If a reverse biased junction diode can give a few microamperes of current when

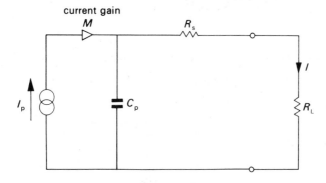

FIGURE 4.14
Equivalent circuit of an avalanche photodiode

irradiated by light, the collector–base junction of a transistor must yield milliamperes under the same conditions, since the photocurrent generated in the reverse biased collector to base junction will be multiplied by h_{fe}, the current gain of the transistor. A typical *npn*-type silicon phototransistor is shown in figures 4.15 and 4.16. Under dark conditions, since the base is not connected to any external circuit, minority carriers are generated thermally, and electrons cross from base to collector while holes drift from collector to base. If I_{c0} is the saturation current of the reverse biased collector to base junction, the external collector current is $I_c = (h_{fe} + 1)I_{c0}$. This is the dark current of the device. If the phototransistor is illuminated with P_{inc}/hf photons per second of incident light, a photocurrent I_p is generated, resulting in total collector current

$$I_c = (h_{fe} + 1)(I_{c0} + I_p)$$

$$I_p = \frac{\eta' P_{inc}}{E_0} \tag{4.19}$$

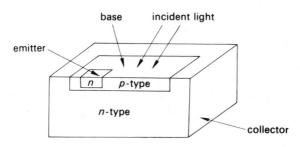

FIGURE 4.15
Geometry of an *npn* phototransistor

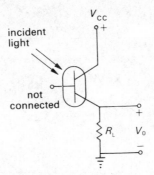

FIGURE 4.16
Floating base bias circuit of a phototransistor

Thus the photocurrent of the reverse biased collector to base junction is multiplied by the large current gain factor ($h_{fe} + 1$).

The collector characteristic of a phototransistor is similar to that of an ordinary transistor except that base current is replaced by light intensity, as shown in figure 4.17. The total incident light collected depends on the active area of the phototransistor. Usually this device comes complete with a lens mounted on the transistor case, giving it a large active area of the order of 0.1 cm^2. The responsivity of the phototransistor of figure 4.17 is about 6 A/W, placing it midway between the PIN photodiode (0.5 A/W) and the avalanche photodiode (60 A/W) in performance.

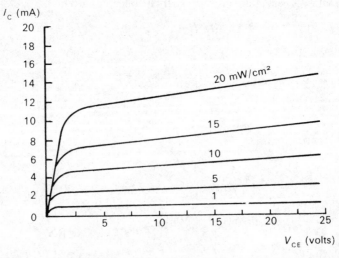

FIGURE 4.17
Collector characteristic of a phototransistor

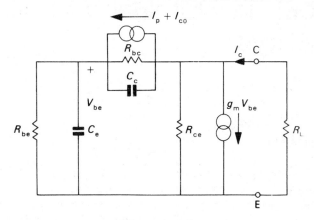

FIGURE 4.18

The hybrid-Π equivalent circuit of a phototransistor used in calculating its frequency response, $h_{fe} = g_m r_{be}$

The frequency response of a phototransistor depends on both the base to collector and base to emitter junction capacitances, and on the effective junction resistances. The junction impedances are shown in the hybrid-Π equivalent circuit of the transistor of figure 4.18. In practice the forward biased emitter–base junction has the lowest resistance and highest capacitance. Consequently the frequency response of the phototransistor is dominated by the time constant $\tau = r_{be} C_e$, which severely limits its bandwidth. Typically $\tau = 1\mu s$ giving a signal bandwidth for the phototransistor of around 160 kHz.

Because the bandwidth of the phototransistor is limited by $r_{be} C_e$ to about 160 kHz, this device is not suitable for use as a detector in a wide bandwidth optical communication system. It finds application in situations where a high current response is needed, such as a sensor of light intensity in the laser power meter and photometer, and in a variety of devices such as the optical paper tape reader.

5 Noise in Optoelectronic Devices

5.1 Introduction

It has long been recognised that a system in thermal equilibrium exhibits internal fluctuations which result from the kinetic energy of molecules and electrons. For example, an isolated resistor at room temperature has internal charge fluctuations which give rise to a randomly fluctuating voltage at its terminals. Even though it may be electrically neutral in the sense that its average voltage is zero, it contains a very large number of atoms whose outer electrons may occasionally gain enough thermal energy to break free and move about, causing local fluctuations in the internal charge density. An electron that has broken free from its particular host atom executes a random walk through the resistor until being recaptured, and its average kinetic energy while free is given by $\frac{1}{2}m\overline{V^2} = \frac{3}{2}kT$, where m is its mass and $\overline{V^2}$ its mean square velocity. Since the mass of an electron is very small its mean square velocity is very large, and thermal fluctuations in charge density in a resistor give rise to a very rapid variation in terminal voltage. This type of noise is often referred to as 'white' noise, since it has a frequency spectrum which is flat out beyond any frequencies of interest in electrical systems. It is more properly called thermal noise.

The dynamics of photons give rise to another sort of noise. Since light is composed of photons, a particular value of light intensity just means an average number of photons per second. Of course, photons are generated randomly, in the same way that electrons undergo random generation and recombination, and therefore the number of photons arriving at a photodetector varies from time to time. If we counted photons over a time interval short enough to detect these fluctuations, then we should find that the number of photons arriving at a photodetector per unit time interval is a random function whose average represents the intensity, and whose standard deviation represents the noise. This type of noise is usually referred to as shot noise, since it is rather like the noise one would hear if a large quantity of shot were thrown against a wall. Shot noise also occurs in semiconductors because of the random generation and recombination of electrons. Each electron thermally generated in a $p-n$ junction, for instance, will drift across the junction until it is recaptured, resulting in a pulse of current. Thus shot noise in a semiconductor gives rise to the super-position of a very large number of such current pulses, which constitute a noise current. The same effect is produced by the random arrival of photons in a photodetector.

Noise is nothing more than a manifestation of the discrete nature of matter. Current in a resistor is the drift of a large number of electrons toward a contact, and is a fluctuating function of time. The mean of these fluctuations is the average current while the standard deviation represents the rms noise current. In practice the two are treated as separate entities, although they are really just two different aspects of the same phenomenon. In the same way local fluctuations in the number of photons gives rise to the shot noise of light. Although noise can be reduced by particular techniques, such as restricting the bandwidth of a communication signal or cooling a photodetector, it can never be eliminated.

Because noise is so fundamental it is often measured in order accurately to determine certain physical constants. For instance, the measurement of the thermal noise of a resistor may be used to calculate Boltzmann's constant. Also, since the performance of most optoelectronic devices cannot be completely specified without reference to their noise characteristics, signal-to-noise ratio and noise equivalent power (NEP) are two methods of specifying the performance of photodetectors.

5.2 Definitions

Before proceeding to discuss various types of noise it is necessary to introduce a few simple ideas which apply to random signals. For a more detailed account of these topics the student should consult one of the numerous texts on probability and stochastic processes. By a random signal we mean one which is continuous in time, but whose shape undergoes rapid and unpredictable variations. If we examine any two segments of a random signal we should find little or no similarity. Furthermore, we shall assume that our signals go on indefinitely, or at least last longer than any times of practical interest. This last assumption leads to a fundamental difficulty. We are used to working with signals and computing their Fourier spectrum

$$F(\omega) = \int_{-\infty}^{\infty} v(t) \exp(j\omega t) \, dt \qquad \text{(Fourier transform of } v(t)) \quad (5.1)$$

However, this expression is not strictly correct unless the signal is integrable

$$\int_{-\infty}^{\infty} | v(t) | \, dt < \infty \qquad \text{(condition for Fourier transform to exist)} \quad (5.2)$$

A random signal which goes on indefinitely does not satisfy this condition of integrability, and therefore its Fourier spectrum does not exist.

Another example of a signal whose Fourier spectrum leads to difficulties is the periodic signal. If a periodic signal is represented by its Fourier series

$$v_p(t) = \sum_{-\infty}^{\infty} V_n \exp(jn\omega_0 t)$$

($\omega_0 = 2\pi/T$, T = the period) then the Fourier transform of this signal is

$$F_p(\omega) = \int_{-\infty}^{\infty} v_p(t) \exp(-j\omega t)\, dt = 2\pi \sum_{-\infty}^{\infty} V_n \delta(\omega - n\omega_0)$$

Because in this case the periodic signal is not integrable in the sense of equation 5.2, its Fourier spectrum contains an infinite number of impulses with infinite height. In order to avoid these difficulties, we usually think of the spectrum of a periodic signal as consisting of a series of lines whose heights are just the Fourier coefficients V_n. Strictly speaking the only signals whose Fourier transforms exist are those of transitory nature, although we commonly apply Fourier analysis to all signals.

It would be a minor disaster to be forced to abandon Fourier analysis in treating random signals, since we want to consider the effects of circuits on these signals, and the circuits will have transfer functions in the frequency domain as usual. This difficulty is overcome by the use of two new ideas, the autocorrelation and power spectrum of a random signal (see figure 5.1).

For the time average of a signal we write

$$\bar{v} = \lim_{T \to \infty} \frac{1}{2T} \int_{-T}^{T} v(\tau)\, d\tau \qquad \text{(d.c. or average signal)} \qquad (5.3)$$

This is just the average, or d.c., value of $v(t)$. The mean square value of a signal is

$$\overline{v^2} = \lim_{T \to \infty} \frac{1}{2T} \int_{-T}^{T} v^2(\tau)\, d\tau \qquad \text{(mean square signal)} \qquad (5.4)$$

The signal rms value is computed from $\overline{v^2}$

$$v_{rms} = \sqrt{(\overline{v^2})} \qquad \text{(rms signal)} \qquad (5.5)$$

The autocorrelation of a random signal $v(t)$ is

$$C(t) = \lim_{T \to \infty} \frac{1}{2T} \int_{-T}^{T} v(t + \tau)v(\tau)\, d\tau \qquad (5.6)$$

In this definition the signal must be convolved with itself and averaged over time. Note that $C(0) = \overline{v^2}$.

The interesting property of the autocorrelation of a signal is that it gradually dies away in time, even though the signal may go on indefinitely. One commonly says that a signal is correlated over a time, τ_c, which is usually called the correlation time. In almost all cases of practical interest, the autocorrelation is integrable, and therefore its Fourier transform exists. The Fourier transform of the autocorrelation is referred to as the power spectrum

$$S(\omega) = \int_{-\infty}^{\infty} C(t) \exp(-j\omega t)\, dt \qquad \text{(power spectrum of } v(t)) \qquad (5.7)$$

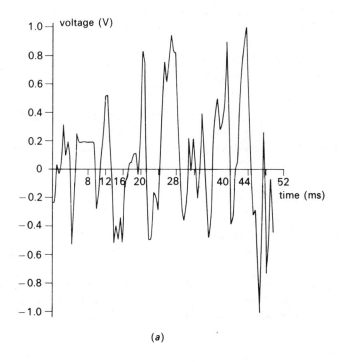

(a)

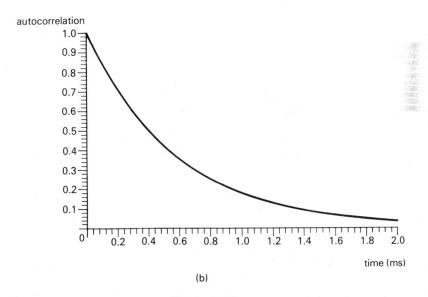

(b)

FIGURE 5.1
(a) A random signal and (b) its autocorrelation

The power spectrum obviously bears some relationship to the Fourier transform of the signal, but is a much more useful idea, since the power spectrum exists even though the signal's Fourier transform may not. Finally, note that using the inverse Fourier transform the autocorrelation of a signal may be computed from its power spectrum.

$$C(t) = \frac{1}{2\pi} \int_{-\infty}^{\infty} S(\omega) \exp(j\omega t) \, d\omega = \int_{-\infty}^{\infty} S(f) \exp(j2\pi ft) \, df \qquad (5.8)$$

A formula of particular interest is that the mean square value of a random signal is given by the integral of its power spectrum

$$\overline{v^2} = C(0) = \int_{-\infty}^{\infty} S(f) \, df \qquad \text{(mean square signal)} \qquad (5.9)$$

One final idea is to investigate what happens when a random signal is passed through a linear circuit, as in figures 5.2 and 5.3. Since the Fourier transform of the input signal does not exist, we cannot compute the output from the inverse transform of the product of $H(\omega)$ and the Fourier spectrum of $v_i(t)$ in the usual way. It may be shown, with a certain amount of difficulty, that the power spectrum of the output is simply related to the power spectrum of the input

$$S_o(\omega) = |H(\omega)|^2 S_i(\omega) \qquad (5.10)$$

where S_o and S_i are the power spectra of the output and input signals, $v_o(t)$ and $v_i(t)$. This relationship will be useful in computing the noise at the output terminals of a circuit from its transfer function.

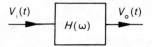

FIGURE 5.2
A random signal applied to a linear circuit with transfer function $H(\omega)$

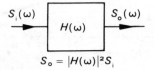

FIGURE 5.3
The transformation of the power spectrum of a random signal through a linear circuit

5.3 Thermal Noise

Thermal noise is illustrated by the random fluctuations in voltage which appear across the terminals of a resistor. These fluctuations arise from the random motion of electrons in the resistor, which at temperature T have an average kinetic energy given by $\frac{1}{2}m\overline{V^2}$ equals $\frac{3}{2}kT$, since the velocity of the electron represents three degrees of freedom. At room temperature the rms velocity of electrons in the resistor is $\sqrt{(3kT/m)}$ or 116 800 m/s. Thus, the voltage fluctuations due to the thermal motion of electrons in a resistor are extremely rapid, and may be approximated as having a flat power spectrum.

In the absence of any external source, the average voltage across the resistor must be zero. If the resistor's noise voltage is $v_n(t)$, then $\overline{v}_n = 0$, and its power spectrum $S_n(\omega)$ may be approximated as a constant, $S_n(\omega) = S_n = $ constant.

When thermal noise is measured it is necessary to consider the bandwidth of the measurement system, since the noise itself has infinite (or very large) bandwidth. One way to specify the thermal noise of a resistor is to consider it to be in thermal equilibrium with a capacitor, as in figures 5.4 and 5.5.

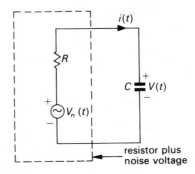

FIGURE 5.4
A resistor and capacitor in thermal equilibrium

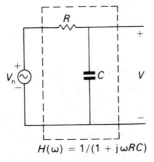

$$H(\omega) = 1/(1 + j\omega RC)$$

FIGURE 5.5
Transfer function of an $R-C$ network

The noise voltage of the resistor causes a current $i(t)$ to flow through the capacitor, and the resulting capacitor voltage, $v(t)$, is itself a random signal. Since the transfer function between input voltage and output voltage is

$$H(\omega) = \frac{1}{(1 + j\omega RC)} \tag{5.11}$$

the power spectrum of the capacitor voltage is

$$S_c(\omega) = |H(\omega)|^2 S_n = \frac{S_n}{(1 + \omega^2 R^2 C^2)} \tag{5.12}$$

The average energy stored in the capacitor is $\frac{1}{2}C\overline{v^2}$, which may be computed by recalling that the integral of the power spectrum of the capacitor voltage is the mean square capacitor voltage $\overline{v^2}$ (see equation 5.9). Thus

$$\overline{v^2} = \int_{-\infty}^{\infty} \frac{S_n}{1 + (2\pi f)^2 R^2 C^2} \, df \tag{5.13}$$

The substitution of $2\pi fRC = \tan\theta$ gives

$$\overline{v^2} = \frac{2S_n}{2\pi RC} \int_0^{\pi/2} d\theta = \frac{S_n}{2RC}$$

The average energy stored in the capacitor is therefore

$$\tfrac{1}{2}C\overline{v^2} = \frac{S_n}{4R} = \tfrac{1}{2}kT \tag{5.14}$$

Since the average thermal energy of the capacitor is $\frac{1}{2}kT$, we can conclude that the power spectrum of thermal noise in the resistor R is

$$S_n = 2kTR \qquad \text{(power spectrum of thermal noise in } R) \tag{5.15}$$

Note also that the power spectrum in this case has units of V^2 s or V^2 Hz^{-1}.

A resistor in a circuit with bandwidth, B, contributes a mean square noise voltage given by

$$\overline{v_n^2} = \int_{-B}^{B} S_n \, df = 2BS_n = 4kTRB \tag{5.16}$$

For example, at room temperature a 1 MΩ resistor in a stereo amplifier, with $B = 30$ kHz, contributes an rms noise voltage equal to 22.3 μV, while the same resistor in a video amplifier with $B = 10$ MHz, would contribute an rms noise voltage equal to 407 μV. The mean square noise per unit bandwidth for a 1 MΩ resistor is 1.656×10^{-14} V^2 Hz^{-1}. We may conclude that in a stereo amplifier it is pointless to try to amplify signal voltages smaller than a few tens of microvolts, and that a video amplifier can only amplify voltages of the order of millivolts or larger.

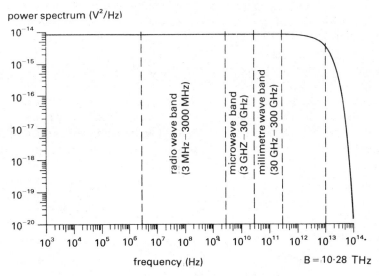

FIGURE 5.6
The power spectrum of thermal noise at room temperature

5.4 The Bandwidth of Thermal Noise*

If it were really true that the power spectrum of thermal noise was flat at all
frequencies, the mean square voltage across a resistor, due to thermal fluctuations,
would be infinite, since in equation 5.16 if $B \to \infty$ so does $\overline{v_n^2}$. This difficulty
was recognised, in slightly different context, by the physicist Planck who derived
the 'blackbody radiation' formula. In a resistor the thermal energy is shared
among the atoms and electrons equally. For instance, for each vibrational mode
of an atom in a resistor there is an average kinetic energy equal to $kT/2$. In our
discussion of thermal noise, we have tacitly assumed that the atoms can have any
amount of kinetic energy provided the average is $kT/2$ per degree of freedom.
Of course, we know that this is incorrect — the energy of atoms occurs only in
discrete levels, as dictated by quantum mechanics. Planck assumed that the
atoms could have only discrete energies, which were proportional to an unknown
constant h. He then showed that the assumption of discrete energies led to a
power spectrum for thermal noise in a resistor given by

$$S_n(\omega) = 2kTR \; \frac{hf/kT}{[\exp(hf/kT)] - 1} \qquad \text{(Planck's thermal noise spectrum)}$$

$$(5.17)$$

---

* This section is included for historical interest, and may be omitted by the reader.

The constant, h, called Planck's constant, was subsequently shown to be equal to 6.626×10^{-34} J s. It is now accepted that transitions between the discrete energy levels of atoms either release or absorb photons, of frequency f and energy hf.

Planck's thermal noise spectrum neatly avoids the infinite mean square noise voltage. At low frequencies where $hf \ll kT$, $S_n = 2kTR$, while at very high frequencies $S_n \rightarrow 2hfR \exp(-hf/kT) \rightarrow 0$. The bandwidth of thermal noise at room temperature is $B \equiv (kT/h)\pi^2/6 = 10\ 280$ GHz. Thus for any frequency of practical interest in an electronic circuit the thermal noise spectrum is flat.

Using equation 5.17, the total mean square noise voltage of a resistor may be computed. Thus

$$\overline{v_n^2} = 2 \int_0^\infty S_n(f)\,df = 4kTR \int_0^\infty \frac{hf/kT}{[\exp(hf/kT)] - 1}\,df \quad (5.18)$$

The substitution of $x = hf/kT$ gives

$$\overline{v_n^2} = 4kTR \left(\frac{kT}{h}\right) \int_0^\infty \frac{x\,dx}{\exp(x) - 1}$$

The value of this integral is $\pi^2/6$. Thus the mean square thermal noise is

$$\overline{v_n^2} = 4kTR \left(\frac{kT}{h}\right)\left(\frac{\pi^2}{6}\right) \qquad (5.19)$$

From this equation we may conclude that the effective bandwidth of thermal noise is $B = (kT/h)(\pi^2/6)$ Hz.

The total rms noise voltage of a 1 MΩ resistor at room temperature is $4kTRB = 0.413$ V, about half a volt spread over an enormous bandwidth! It is, in principle, possible to extract electrical power from thermal noise, although the total amount available from a 1 MΩ resistor at room temperature is rather small, $0.17\ \mu$W.

5.5 Noise Sources in Circuits

In the circuit representation of noise sources, each resistor contributes a random voltage with mean square voltage $\overline{v_n^2}$. It is sometimes convenient to work with a current source rather than a voltage source. The pair of equivalent thermal noise representations is shown in figure 5.7. The mean square current source is $\overline{i_n^2} = \overline{v_n^2}/R^2$.

Two resistors in series contribute a noise voltage equal to that of their combined resistances. This can be shown by considering the sum of the two random

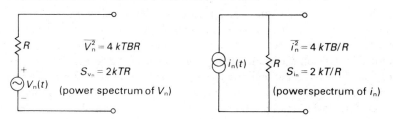

FIGURE 5.7

The equivalent voltage and current sources for a thermal noise source R

voltages $v_n = v_1 + v_2$. The mean square value of the sum is

$$\overline{v_n^2} = \overline{(v_1 + v_2)^2} = \overline{v_1^2} + 2\overline{v_1 v_2} + \overline{v_2^2}$$

But the average of the product $\overline{v_1 v_2}$ is zero since v_1 and v_2 are statistically independent. Thus

$$\overline{v_n^2} = \overline{(v_1 + v_2)^2} = \overline{v_1^2} + \overline{v_2^2} = 4kTB(R_1 + R_2) \qquad (5.20)$$

In more complicated situations the noise sources may be determined by first combining all resistors to simplify the circuit. The power spectrum of the resulting noise voltage is $2kTR$, where R is the combined circuit resistance (see figure 5.8).

5.6 Shot Noise

The random arrival of photons at a photodetector gives rise to a random fluctuation in photocurrent known as shot noise. The current, if examined over

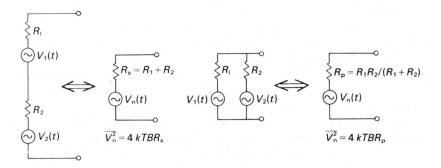

FIGURE 5.8

The combination of two thermal noise sources in: (a) series; and (b) parallel

very short time intervals, consists of a random series of pulses, each pulse corresponding to the absorption of a photon. Photon statistics, and consequently the random release and recombination of electrons in a semiconductor, is governed by Poisson statistics. For example, consider the detection of a 1 mW laser beam by a photodiode. For a He–Ne laser with $\lambda_0 = 632.8$ nm, the average number of photons arriving per second is $\bar{n} = 3.2 \times 10^{15}$ photons per second. The standard deviation in photon arrivals, that is the statistical variation about the average, is $\sigma_n = \sqrt{\bar{n}} = 5.6 \times 10^7$ photons per second. We conclude that such a laser beam carries with it a statistical variation of the order of one part in 10^7, if the intensity is observed over a 1 s time interval. The statistical variation over a 1 μs time interval would be of the order of one part in 10^4, since the average number of photons in 1 μs is 2.7×10^9, while the statistical variation over a 1 ps time interval would be of the order of 1 part in 50.

Depending on the quantum efficiency of the photodetector, absorbed light produces a steady photocurrent I_p, proportional to the incident light power, and a noise current resulting from the discrete arrival times of photons (see figure 5.9). The power spectrum of the shot noise current of a photodetector is

$$S(\omega) = eI_p \qquad \text{(power spectrum of shot noise current} \qquad (5.21)$$
$$\text{resulting from an average photocurrent } I_p)$$

The mean square noise current in bandwidth B is

$$\overline{i_n^2} = 2eI_pB \qquad \text{(mean square shot noise current in bandwidth } B) \qquad (5.22)$$

The total photocurrent is

$$i_p(t) = \underbrace{I_p}_{\substack{\text{average or} \\ \text{d.c. current}}} + \underbrace{i_n(t)}_{\text{noise current}} \qquad (5.23)$$

In the equivalent circuit of a photodiode we should use two current sources, one for the average or d.c. photocurrent and the other for the shot noise current (see figure 5.10).

Practically speaking, there is little difference between shot noise and thermal noise. They are both very wide bandwidth phenomena. However, the fact that they originate from very different physical phenomena – thermal noise from the random *motion* of charges, and shot noise from the random *times of release* of charges – gives rise to one important difference. Whereas thermal noise is a property of the resistors of a circuit, and therefore is subject to some control by the designer, shot noise is an inherent property of photodetectors, and even of the optical signal itself. For this reason, shot noise can never be eliminated. An investigation of most photodetectors will show that shot noise is the dominant feature, especially if we consider that even in the absence of light a photodetector's internal charges are continuously being generated and recombined because of thermal energy in the material.

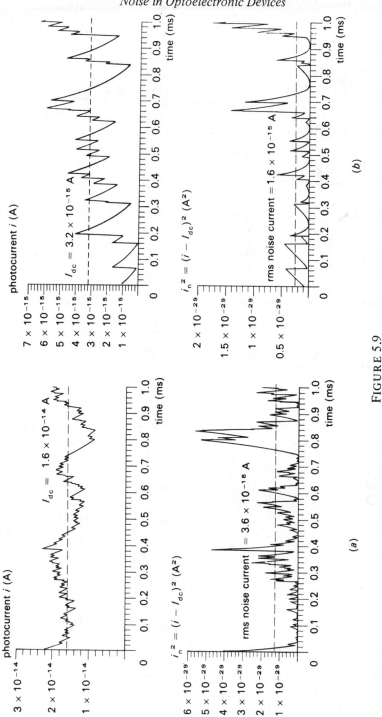

FIGURE 5.9

The random nature of photocurrent resulting from the discrete arrival of photons at a photodetector with response time $\tau = RC$. In (a) the rate is 10^5 pulses per second, while in (b) it is 2×10^4 pulses per second. The resulting d.c. and rms noise currents are also shown. In (a) the rms noise current is 22.4 per cent of the d.c. current, while in (b) it is 50 per cent

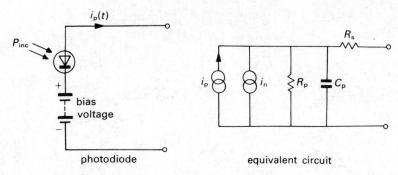

FIGURE 5.10

Equivalent circuit of a photodiode including the shot noise current

5.7   Signal-to-noise Ratio in Photodetectors

Consider first a photodiode connected to a load resistance R_L, which could be the feedback resistance of an operational amplifier, as in figure 5.11. The power spectrum of the shot noise current is $S_{sn} = eI_p$ while that of the thermal noise current is $S_{th} = 2kT/R_L$. Since the two noise sources are statistically independent, we may compute their mean square contribution separately and add the result to obtain the total noise current.

The mean square shot noise current is

$$\overline{i_n^2} = \int_{-\infty}^{\infty} \frac{S_{sn}\,df}{1 + (2\pi fR_L C_p)^2} = \frac{eI_p}{2R_L C_p} = 2eI_p B \tag{5.24}$$

where the effective bandwidth of the circuit is $B = 1/4R_L C_p$. Similarly, the mean square thermal noise current is

$$\overline{i_{n_L}^2} = \left(\frac{4kT}{R_L}\right) B \tag{5.25}$$

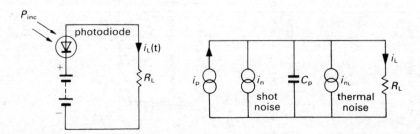

FIGURE 5.11

Thermal and shot noise sources of a photodiode circuit

The total mean square noise current is

$$\overline{(i_n + i_{n_L})^2} = \overline{i_n^2} + \overline{i_{n_L}^2} = \left(2eI_p + \frac{4kT}{R_L}\right)B \qquad (5.26)$$

The signal-to-noise ratio in the load resistance is the ratio of d.c. photocurrent to rms noise current, the square root of the sum of mean square shot and thermal noise currents. The result is

$$\frac{S}{N} = \frac{I_p}{\sqrt{(2eI_pB + 4kTB/R_L)}} \qquad \text{(photodiode)} \qquad (5.27)$$

Suppose that the photocurrent is $I_p = 1$ mA, $C_p = 5$ pF, $R_L = 1$ MΩ and $T = 300$ K. Then $B = 50$ kHz and the mean square shot noise current is 1.602×10^{-17} A^2 while the mean square thermal noise current is 0.828×10^{-21} A^2. The rms noise current is approximately 4 nA and the signal-to-noise ratio is

$$\frac{S}{N} = \frac{1 \text{ mA}}{4 \text{ nA}} = 2.5 \times 10^5 = 108 \text{ db} \qquad \text{(photodiode)}$$

Consider a typical photoconductor circuit where a photocurrent is generated by an external bias voltage, as in figure 5.12. The two sources of noise, shot noise in the photoconductor and thermal noise in the load resistance and photoconductor, give mean square noise current $2eI_pB$ and $4kTB/R_{11}$, where $B = 1/4C_pR_{11}$, and $R_{11} = RR_L/(R + R_L)$.

The signal-to-noise ratio in this circuit is

$$\frac{S}{N} = \frac{I_p}{\sqrt{(2eI_pB + 4kTB/R_{11})}} \qquad \text{(photoconductor)} \qquad (5.28)$$

Typical values are $I_p = 100$ mA, $T = 1$ ms, $R = 1$ kΩ. If the load resistance is $R_L = R = 1$ kΩ, the bandwidth is $B = 500$ Hz. The mean square shot noise current is 1.602×10^{-17}A^2, while the mean square thermal noise current, assuming $T = 300$ K, is 1.656×10^{-20}A^2. The rms noise current is again 4 nA, and the

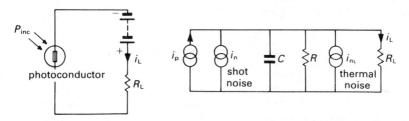

FIGURE 5.12
Thermal and shot noise of a photoconductor circuit

signal-to-noise ratio is

$$\frac{S}{N} = \frac{100 \text{ mA}}{4 \text{ nA}} = 2.5 \times 10^7 = 148 \text{ db} \qquad \text{(photoconductor)}$$

The signal-to-noise ratios calculated here are unrealistic, since the d.c. photocurrent has been treated as the signal. In practice the light incident on the photodetector would be modulated, resulting in a signal photocurrent, $i_s(t)$. The rms value of this signal current would constitute the signal, and in general would be much smaller than the d.c. photocurrent, resulting in smaller S/N than those calculated above. There is no need to elaborate this point, since the reader will be aware that a signal-to-noise ratio of 60 db is considered good in practice. In both the photodiode and photoconductor, the largest contribution to noise is their shot noise.

5.8 Noise Equivalent Power

The noise equivalent power, NEP, of a photodetector is the amount of incident light, measured in watts, which would generate a photocurrent equal to the dark current of the device. In the photodiode, the dark current is the reverse bias saturation current, while in the photoconductor it is the ratio of bias voltage to bulk resistance under dark conditions. Since the dark current is shot noise, the significance of the NEP is that it represents the lowest level of incident light which could be measured. Light of less intensity would result in a photocurrent which would be indistinguishable from the noise current of the photodetector. Thus the NEP is a measure of the sensitivity of the photodetector.

The NEP is usually defined slightly differently, as the rms value of the sinusoidally modulated incident light which will give rise to a rms signal current. equal to the rms noise current per unit bandwidth from the photodetector. In a reverse biased photodiode the mean square shot noise current in bandwidth B which results from dark current I_D is $2eI_D B$, the mean square shot noise per unit bandwidth is $2eI_D$, and the rms shot noise per unit bandwidth is $\sqrt{(2eI_D)}$, with units of A Hz$^{-1/2}$. If the current responsivity of the photodiode is $R_\phi \equiv$ A/W, then the noise equivalent power is

$$\text{NEP} = \frac{\sqrt{(2eI_D)}}{R_\phi} \qquad \text{(photodiode with dark current } I_D) \qquad (5.29)$$

The noise equivalent power has units of W Hz$^{-1/2}$.

It should be realised that the noise equivalent power is wavelength dependent, through the current responsivity R_ϕ, and modulation frequency dependent because of the photodetector's shunt capacitance. It is customary to write NEP(λ, f_m), to indicate these dependencies explicitly.

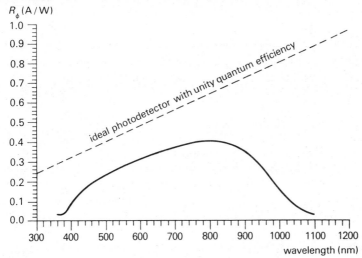

FIGURE 5.13
Current responsivity of a silicon photodiode

As an example, consider a photodiode with dark current I_D = 5 nA (reverse bias voltage = −25 V) and a variation of current responsivity with wavelength as in figure 5.13. The peak current responsivity at 800 nm is 0.5 A/W. Using I_D = 5 nA, $\sqrt{(2eI_D)} = 4 \times 10^{-14}$ A Hz$^{-1/2}$, and NEP(800 nm, f_m) = 8×10^{-14} W Hz$^{-1/2}$. The variation of NEP with wavelength for this photodiode is shown in figure 5.14.

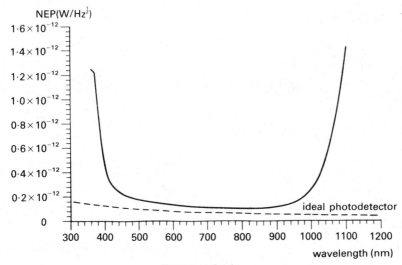

FIGURE 5.14
Variation of the noise equivalent power (NEP) with wavelength

Using NEP = 8×10^{-14} W Hz$^{-1/2}$ in a broadband circuit with B = 1 MΩ, the noise equivalent power in that bandwidth is $8 \times 10^{-14} \times \sqrt{10^6}$ = 8×10^{-11} W, or 80 pW. Thus the sensitivity of this photodiode is quite high. This much power falling on one square millimetre gives a light intensity equal to 8 nW/cm^2, approximately 10^{+7} times smaller than bright sunlight, or 100 times smaller than the intensity of moonlight.

5.9 Evaluation of a Photodiode Preamplifier

Having discussed various types of noise and the different methods of evaluating it, we are in a position to analyse the photodiode operational amplifier circuit of chapter 4. In this circuit the effective noise bandwidth is $B = 1/4R_L C_p$ and the mean square shot noise current is $2eI_p B$, where I_p is the d.c. photocurrent. If the signal current is $i_s(t)$ with rms value $\sqrt{(i_s^2)}$, the signal-to-noise ratio including the thermal noise current from R_L is

$$\frac{S}{N} = \frac{\sqrt{(i_s^2)}}{\sqrt{(2eI_p B + 4kTB/R_L)}} \tag{5.30}$$

The minimum detectable signal current in this circuit (see figure 5.15) may be calculated by equating the mean square signal current to the mean square noise current

$$\overline{(i_s^2)} \text{ min} = 2eI_p B + \frac{4kTB}{R_L} \tag{5.31}$$

Some typical values at room temperature are shown in table 5.1.

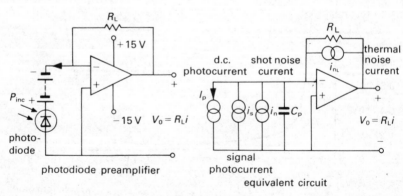

FIGURE 5.15
Photodiode preamplifier and equivalent circuit

TABLE 5.1

DC Photocurrent $I_p(\mu A)$	Shot Noise $2eI_pB$ $(\times 10^{-20})$	Thermal Noise $4kTB/R_L$ $(\times 10^{-20})$	Minimum rms Signal Current (nA)
10.0	32.04	1.134	0.576
1.0	3.024	1.134	0.204
0.1	0.3204	1.134	0.121
0.01	0.03204	1.134	0.108

$B = 100$ kHz, $R_L = 146$ kΩ, $T = 300$ K

The current through the feedback resistor R_L, ignoring the d.c. photocurrent, is

$$i_L(t) = i_s(t) + i_n(t) + i_{th}(t) \qquad (5.32)$$

The power spectrum of the load current, assuming the signal bandwidth is smaller than the preamplifier bandwidth, is

$$S_L(\omega) = S_S(\omega) + \frac{(eI_p + 2kT/R_L)}{(1 + \omega^2 R_L^2 C_p^2)} \qquad (5.33)$$

where $S_S(\omega)$ is the power spectrum of the signal. The autocorrelation of the current $i_L(t)$ is

$$C_L(t) = C_S(t) + 2eI_pB + \frac{4kTB}{R_L} \exp(-t/R_L C_p) \qquad (5.34)$$

Note that the finite bandwidth of the preamplifier causes an autocorrelation of the form $\exp(-t/R_L C_p)$ for the noise current (see figure 5.16).

As an example of the performance of a photodiode preamplifier used in practice, consider the circuit of figure 5.17. In this circuit the photodiode, operated in the photovoltaic mode, was connected to an FET operational amplifier with $R_L = 146$ kΩ. The circuit includes temperature compensation through the resistor $R_T = 560$ Ω and d.c. offset adjustment through the 10 kΩ multiturn potentiometer. The preamplifier was adjusted so that the d.c. offset on the output was zero when the photocurrent was zero. The preamplifier noise bandwidth was $B = 100$ kHz with $R_L C_p = 2.5$ μs. In a typical example the d.c. photocurrent was $I_p = 1.096$ μA and the rms signal current was $\sqrt{(i_s^2)} = 81.2$ nA. The calculated rms noise current was $\sqrt{(i_n^2)} = 0.19$ nA, which consisted of shot noise current due to the d.c. photocurrent and thermal noise current from the

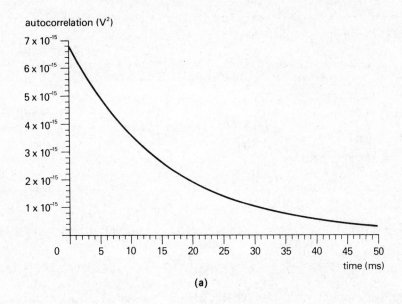

(a)

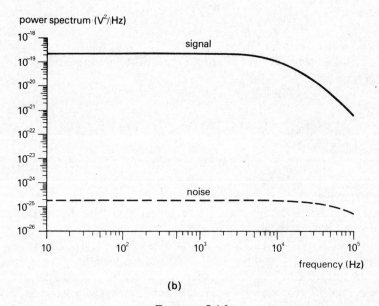

(b)

FIGURE 5.16

(a) The autocorrelation and (b) power spectrum of load current in a photodiode
preamplifier

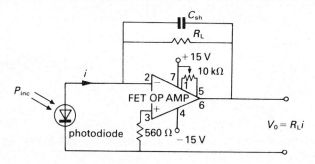

FIGURE 5.17

Details of a photodiode preamplifier with temperature compensation and d.c. offset adjustment. The feedback capacitance prevents oscillation at high signal levels

feedback resistance of the preamplifier. The signal-to-noise ratio was $S/N = 52.6$ db, and the signal plus noise power spectrum and autocorrelation are shown in figure 5.17.

6 The Solar Cell

6.1 Introduction

The technology of solar cells has been developed as a result of the need for long-lasting power supplies for satellites and space vehicles. The device is based on the photovoltaic effect in the photodiode. With no external bias a photodiode will deliver power to a load resistance if irradiated by light with photons of sufficient energy. By making the collection area as large as possible, the solar cell will deliver a large photocurrent from direct illumination by sunlight. A solar cell with a collection area of 30 cm^2 and a power conversion efficiency of 10 per cent will deliver 300 mW to a load if illuminated by bright sunlight.

At the present time, solar cells have not come into widespread use, primarily because the cost is too high. A cost of around £8 per watt is typical. However, it is likely that the cost will diminish with mass production of solar cells for domestic use. They offer many advantages over other methods of converting energy into electrical power, including simplicity, portability and environmental compatibility. The solar cell produces no gaseous or thermal pollutants. Even countries outside the equatorial zone experience a large total daily amount of solar energy. In fact, northern latitudes have a larger total daily amount of sunlight in summer, due to their longer daylight hours, than do equatorial zones. The solar intensity on a bright summer day may be as high as 1 kW/m^2. With a conversion efficiency of 10 per cent, the roof of the average home offers enough surface area for the generation of several hundred watts of electrical power during daylight hours, which could be stored and used throughout the 24 hour cycle. Solar cells may find application in developing parts of the world where it is undesirable or impractical to lay electric cable. They also may be used for the power source in the irrigation of desert areas which enjoy high levels of sunlight.

Although many different semiconductor materials are suitable for the fabrication of solar cells, the silicon solar cell is the only type commercially available. Silicon gives probably the highest power conversion efficiency of all materials tried, and is the most widely available and best understood of all semiconductors.

6.2 The n on p Silicon Photovoltaic Cell

A typical silicon photovoltaic cell, or solar cell, consists of a disc of p-type silicon with a thin n-layer diffused on its top surface (see figure 6.1). The collecting surface is provided with stripe and finger contacts, with a solid metal contact on the back surface. Incident photons pass into the depletion region

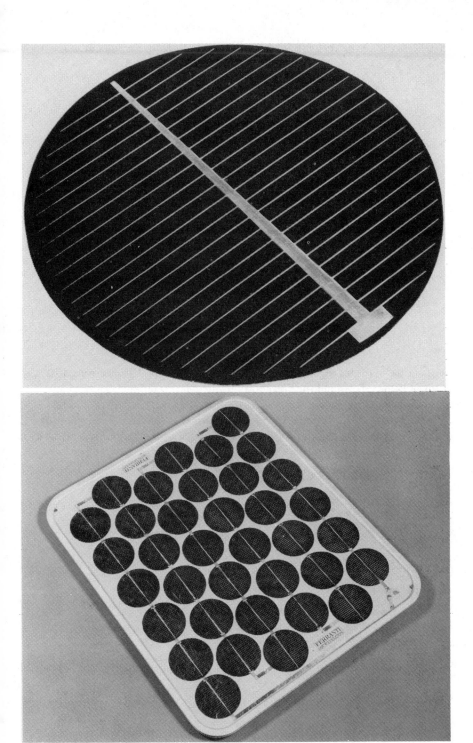

FIGURE 6.1
(top) Solar cell; (bottom) solar module (courtesy of Ferranti Electronic
Components Division)

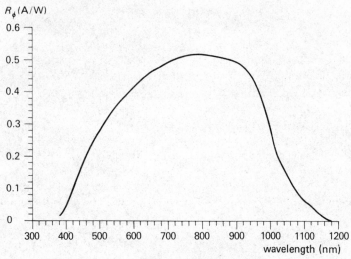

FIGURE 6.2
The current responsivity of a silicon solar cell

where they are absorbed, producing electron–hole pairs. The electrons drift towards the *n*-layer and the holes towards the *p*-layer, producing a reverse photocurrent.

Since in normal operation the photovoltaic cell is forward biased, some of the photocurrent is shunted by the forward biased *p*–*n* junction. The effect of

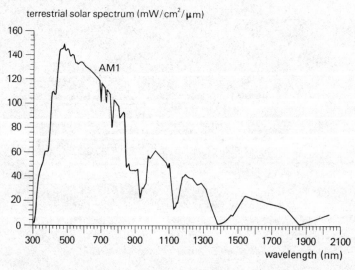

FIGURE 6.3
Spectrum of solar radiation on the surface of the earth (AM1)

the series resistance of the cell, which will be considered later, is to increase the junction voltage for a given cell voltage and to reduce the power through internal power dissipation.

The other factor which influences the power conversion efficiency of the photovoltaic cell is its variation of quantum efficiency with wavelength. The peak quantum efficiency of a silicon solar cell occurs at about 800 nm. The Sun has an emission spectrum equivalent to that of a blackbody at a temperature of 6000 K, centered at a wavelength of 600 nm (see figure 6.3). The solar intensity on the surface of the Earth is also affected by atmospheric absorption, which causes a reduction due to the absorption of different portions of the spectrum by the different molecules of the atmosphere.

The total responsivity of the photovoltaic cell is the integral of the product of current responsivity and solar intensity at each wavelength. The current responsivity of a solar cell is also reduced by surface reflection of incident light.

6.3 Current–Voltage Characteristics

The solar cell is a p–n junction diode with reverse photocurrent generated by absorbed light. The cell current is

$$I = I_p - I_s \; [\exp(eV/kT) - 1] \tag{6.1}$$

where I_p is the photocurrent, and the second term is the forward diode current, which is significant since the cell voltage is positive (see figure 6.4).

The short circuit current of the solar cell is

$$I_{sc} = I_p \tag{6.2}$$

while the open circuit voltage is given by

$$V_{oc} = \left(\frac{kT}{e}\right) \ln\left(1 + \frac{I_p}{I_s}\right) \tag{6.3}$$

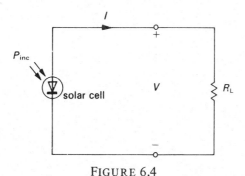

FIGURE 6.4
A solar cell and load resistance

At $T = 300$ K, kT/e equals approximately 25.8 mV.

The maximum power obtainable from the solar cell is slightly less than the product of short circuit current and open circuit voltage

$$P_{\max} < I_{sc}V_{oc} = V_o I_p \ln\left(1 + \frac{I_p}{I_s}\right) \qquad (6.4)$$

where $V_o = kT/e$. It is obvious from this expression that a high efficiency solar cell must produce as large a photocurrent as possible for a given surface area, and must have as small a reverse saturation current, I_s, as possible.

A plot of the current–voltage characteristics of an ideal solar cell is shown in figure 6.5. The output power from the cell is just the product of voltage and current at a given point on the characteristic curve. It is obvious that the output power is always less than the product of short circuit current and open circuit voltage

$$P_{out} = IV = I_p V - I_s V \left[\exp\left(\frac{V}{V_0}\right) - 1\right] \qquad (6.5)$$

The maximum power occurs at the point where $\partial P_{out}/\partial V = 0$. Thus

$$\frac{\partial P_{out}}{\partial V} = I_p - I_s\left[\exp\left(\frac{V}{V_0}\right) - 1\right] - I_s\left(\frac{V}{V_0}\right)\exp\left(\frac{V}{V_0}\right) = 0$$

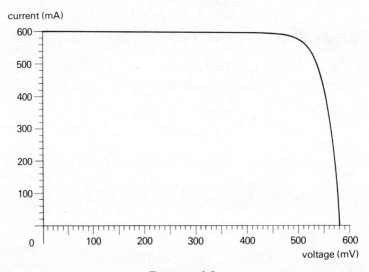

FIGURE 6.5
Current–voltage characteristics of a solar cell with zero series resistance

which gives

$$\left(1 + \frac{V}{V_o}\right) \exp\left(\frac{V}{V_o}\right) = 1 + \frac{I_p}{I_s} \tag{6.6}$$

This transcendental equation may be solved for the voltage which gives maximum power provided I_p and I_s are known. Note that in the solar cell represented by figure 6.5, the photocurrent is $I_p = 600$ mA while the open circuit voltage is $V_{oc} = 580$ mV. Thus, in this example with zero series resistance

$$1 + \frac{I_p}{I_s} = \exp\left(\frac{V_{oc}}{V_o}\right) = \exp\left(\frac{580}{25.8}\right)$$

and the condition for maximum power is

$$\left(1 + \frac{V}{V_o}\right) \exp\left(\frac{V}{V_o}\right) = \exp\left(\frac{580}{25.8}\right)$$

The solution is $V/V_o = 19.5$ or $V = 502$ mV. The current at this voltage is $I = 572$ mA and the maximum power is $P_{max} = 287$ mW. Since the product of short circuit current and open circuit voltage is 348 mW, the maximum power is within 82 per cent of $I_{sc}V_{oc}$. The load resistance which gives maximum power in this example is

$$R_L = \frac{V}{I} = \frac{502\,\text{mV}}{572\,\text{mA}} = 0.88\,\Omega$$

In this discussion of the ideal solar cell we have neglected one important effect, the series resistance of the cell. If the $p-n$ junction has a small series resistance, say $R_s = 1.0\,\Omega$, the power dissipated within the cell with a current of 572 mA would be 327 mW, nearly as large as the maximum power generated by the ideal solar cell!

6.4   Series Resistance Loss

The model of a solar cell with series resistance is the ideal diode in series with R_s, as in figure 6.6. It is obvious from this diagram that the junction voltage no longer equals the cell voltage. If the cell current is $I = I_p - I_s[\exp(V_d/V_o) - 1]$, the junction voltage is

$$V_d = V + IR_s \tag{6.7}$$

For a given cell voltage the junction voltage is higher then in the ideal solar cell, and therefore the cell current is reduced.

The relationship between cell voltage and current, incorporating series

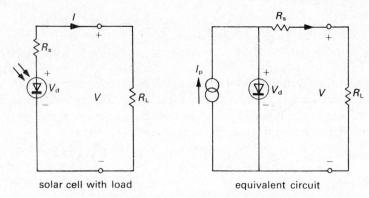

solar cell with load   equivalent circuit

FIGURE 6.6

The solar cell with series resistance, R_s

resistance R_s, is

$$V = -IR_s + V_o \ln\left[1 + \frac{(I_p - I)}{I_s}\right]$$ (6.8)

This expression shows that for a given cell current, the cell voltage is reduced by the voltage drop across the series resistance of the solar cell.

A plot of the current–voltage characteristics of a solar cell with different values of series resistance is shown in figure 6.7, using the same parameters as in

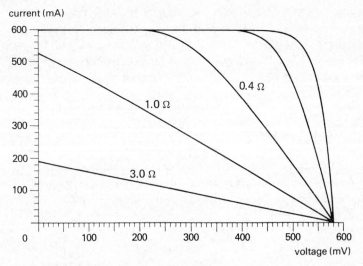

FIGURE 6.7

Current–voltage characteristics of a solar cell with $R_s = 0, 0.11, 0.4, 1.0$, and $3.0\ \Omega$

figure 6.5. It may be concluded from this diagram that even a series resistance as small as 0.11 Ω has a significant effect on the performance of the cell.

In order to study the effect of series resistance on the power output of the solar cell, the maximum power was calculated for each of the curves of figure 6.7 and plotted against R_s. If the cell has $V_{oc} = 580$ mV and $I_p = 600$ mA, as before, the maximum output with $R_s = 0.11$ Ω is

$$P_{max} = 251 \text{ mW} \qquad (6.9)$$

This is about 87 per cent of the power obtainable from the ideal cell, while with $R_s = 0.4$ Ω the power drops to 167 mW, 58 per cent of that obtainable from the ideal cell. We see that it is desirable to fabricate solar cells with series resistance loss of the order of 0.1 Ω or smaller.

6.5   Spectral Characteristics of the Silicon Solar Cell

Provided a solar cell may be fabricated with small series resistance (see figure 6.8), the factor which determines its output power is the photocurrent. We have seen in chapter 4 that the photocurrent which results from a total incident light of P_{inc} W at wavelength λ is

$$I_p(\lambda) = \frac{\eta(\lambda)P_{inc}(\lambda)}{E_0} \qquad (6.10)$$

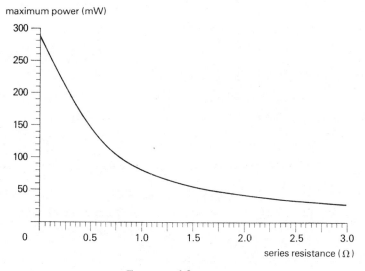

FIGURE 6.8
Solar cell power output versus series resistance

where $E_0 = hf/e = hc/\lambda e$ is the energy in electron volts of incident photons and $\eta(\lambda)$ is the quantum efficiency. The quantum efficiency depends on such factors as the absorption coefficient and thickness of the depletion region of the $p-n$ junction. Another important consideration in quantum efficiency, which has not been previously mentioned, is the reflection loss at the air–solar-cell boundary.

At a wavelength of 632.8 nm in the visible spectrum, the refractive index of silicon is approximately equal to 3.75. The intensity reflection coefficient, the ratio of reflected to incident light intensity, assuming normal incidence is

$$R = \left(\frac{n-1}{n+1}\right)^2 = \left(\frac{3.75-1}{3.75+1}\right)^2 = 0.34 \qquad (6.11)$$

Thus approximately 34 per cent of the incident photons are reflected before having the chance to enter the junction depletion region and be absorbed.

If the light is not normally incident, an even greater amount is reflected from the collecting surface of the solar cell (see figure 6.9). For optimum performance of the cell, it is necessary to point it directly at the Sun at all times. However, the amount of sunlight reaching Earth's surface during morning and evening hours is considerably reduced by the longer transmission path through the atmosphere, so that the effect of reflection loss over a 24 hour period may not be as important as might be expected.

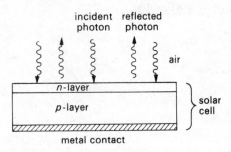

FIGURE 6.9

Reflection loss at the surface of a solar cell

The reflection loss with normal incidence predicted by equation 6.11 may be larger than the real figure, since the solar cell consists of a thin n-layer on top of a thick p-layer. The n-layer acts as an antireflection coating, reducing the intensity reflection coefficient. The peak quantum efficiency of a silicon photodiode is about 0.8, indicating a reflection of about 20 per cent.

The current responsivity of the solar cell at wavelength λ is

$$R_\phi(\lambda) = \frac{I_p(\lambda)}{P_{inc}(\lambda)} = \frac{\eta(\lambda)}{E_0} \qquad (6.12)$$

The current responsivity of a typical silicon solar cell is shown in figure 6.2. The peak response occurs at about 800 nm. The spectral response of this type of solar cell spans the visible and a portion of the infrared spectrum. The following section discusses the spectrum of solar radiation reaching the Earth and shows the calculation of an actual photocurrent of a silicon solar cell.

6.6   Spectrum of Solar Radiation and Atmosphere Absorption

The Sun is a very large mass of hot gases, composed primarily of hydrogen and helium, which emits strong radiation in the visible portion of the spectrum. Although a certain amount of absorption by the Sun of its own radiation occurs in the ultraviolet and violet end of the spectrum, in the infrared portion its emission spectrum closely resembles that of blackbody at a temperature of 6000 K. The total intensity of the Sun's radiation above the atmosphere – the extraterrestrial solar constant – is 135.3 mW/cm^2, with peak emission at a wavelength of about 480 nm. Sunlight is the primary source of energy on Earth, responsible through plant life and photosynthesis for the diminishing deposits of coal and other fuels, as well as for the maintenance of a suitable terrestrial environment for all life. The Sun has been shining for 500 million years and is expected to continue to do so for another 50 million years.

As sunlight travels through the atmosphere it is absorbed and scattered by the various gases, dust particles, and water vapour molecules. Of the scattered light, about half is directed toward the surface of the Earth. It is estimated that only 6 per cent of the incident sunlight is back-scattered away from the earth by the atmosphere, 4 per cent is absorbed in the stratosphere and 17 per cent absorbed in the troposphere. However, thick cloud can reflect as much as 27 per cent of the incident sunlight. The strongest absorption in the atmosphere is by ozone, which absorbs the ultraviolet portion of the spectrum. Carbon dioxide, water vapour and ozone are also responsible for absorption bands in the near infrared portion of the spectrum. On a cloudless day about 73 per cent of the incident sunlight reaches the surface of the Earth (assuming the Sun is directly overhead), amounting to a power density of 88.92 mW/cm^2.

As the Sun moves from east to west, the thickness of atmosphere which the sunlight must penetrate to reach the surface of the Earth changes. This effect is usually expressed in terms of the optical air mass m. The air mass is defined as $m = \sec z$, where z is the angle between a vertical line and a line through the observer and the Sun. Air mass one (AM1) indicates that the Sun is directly overhead while air mass zero (AM0) is used to indicate extraterrestrial sunlight (see figure 6.10). The spectrum of sunlight reaching the surface of the Earth (AM1) is shown in figure 6.11. The atmospheric absorption bands in the near infrared are clearly evident. The solar spectrum for air mass four (AM4), when the Sun is nearer the horizon, is also shown.

Since the effect of cloud is to reflect, rather than absorb sunlight, a broken

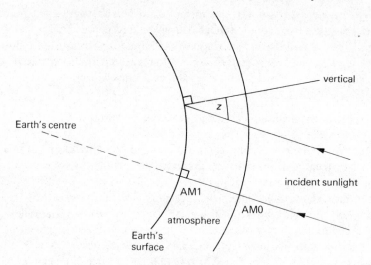

FIGURE 6.10
The definition of the optical air mass

cloud cover can produce a higher solar intensity on the surface of the Earth under certain circumstances. Although the daily average intensity is about the same, higher peak intensities are reached on days with well broken cloud.

Another interesting terrestrial phenomenon is the variation of total daily sunlight on the surface of the Earth at different latitudes. Although the

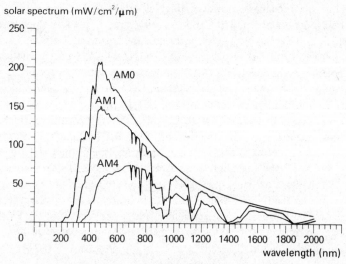

FIGURE 6.11
The extraterrestrial (AM0) and terrestrial (AM1 and AM4) solar spectrum

equatorial zones have the brightest noontime sunlight, the longer days in summer in northerly (or southerly) latitudes give a larger total sunlight. In fact, the largest total daily sunlight occurs at both poles in their respective summers, provided that cloud cover does not prevent direct sunlight reaching the surface of the Earth.

6.7   Photocurrent and Power Conversion Efficiency

The photocurrent which results from the absorption of sunlight by a solar cell may be calculated by integrating the product of current responsivity and solar intensity over all wavelengths (see equation 6.10 and figure 6.12). The product of the current responsivity of a silicon solar cell (figure 6.2) and the solar spectrum for AM0, AM1, and AM4 conditions (figure 6.11) is shown in figure 6.13. The area under each of the three curves gives the current per unit solar cell area, which is 33.0 mA/cm^2 (AM0), 24.4 mA/cm^2 (AM1) and 13.0 mA/cm^2 (AM4). If the solar cell has a collection area of 24.6 cm^2, the total photocurrent is 811 mA (AM0), 600 mA (AM1) and 319 mA (AM4). The peak quantum efficiency is 0.807 electrons per photon. Although some improvement in quantum efficiency might be possible, the resulting change in photocurrent would at best amount to a 20 per cent increase.

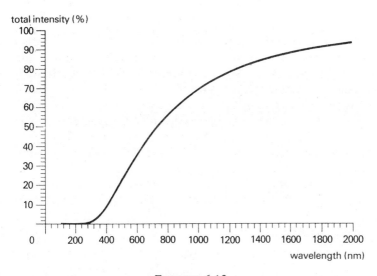

FIGURE 6.12

The integrated solar spectrum showing the proportion of the total intensity (135.3 mW/cm^2) contained in the wavelength range from zero to λ

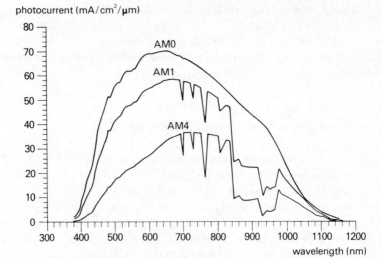

FIGURE 6.13

The product of current responsivity of a silicon solar cell x the solar spectra of
figure 6.11

Since the photocurrent generated is given by

$$I_p = \int_0^\infty R_\phi(\lambda) P_{inc}(\lambda) \, d\lambda \qquad (6.13)$$

a solar cell with a current responsivity of $R_\phi = 0.5$ A/W at all wavelengths (the
peak current responsivity of a silicon solar cell) would give a photocurrent equal
to 0.5 x 135.3 x 24.6 = 1.664 A, under AM0 conditions. This is approximately
twice as large as the value of 811 mA calculated above. Due to the spectral
response of the cell, its average current responsivity is about half its peak
responsivity. If it was possible to improve the spectral response the photocurrent
could be nearly doubled. We shall return to this point in the next section.

The power conversion efficiency of the solar cell is defined as the ratio of
electrical power output to total incident solar power. Referring to section 6.4,
the maximum power obtainable from the n on p silicon solar cell under AM1
conditions, with A = 24.6 cm^2 and I_p = 600 mA, is 287 mW. The incident solar
power is 88.92 mW/cm^2 (AM1) times 24.6 cm^2, which equals 2.187 W. The
power conversion efficiency is

$$\eta_{max} = \frac{P_{max}}{P_{inc}} = \frac{287 \text{ mW}}{2.187 \text{ W}} = 13.1 \% \qquad (6.14)$$

We see that approximately 13 per cent of the incident solar power is converted
to electrical power. A small series resistance will reduce the efficiency even more.

The reasons for this rather poor conversion efficiency will be discussed in the next section.

6.8 Can Solar Cell Efficiency be Increased?

The power conversion efficiency of a solar cell is affected by several different factors. The cell's spectral response determines its photocurrent while the reverse bias saturation current, I_s, together with the photocurrent determine its open circuit voltage. These two factors, I_p and V_{oc}, determine the maximum output. If the cell has appreciable series resistance, it will dissipate power internally reducing the power output even further.

Assuming that the cell series resistance can be reduced by proper design and doping of silicon, consider a cell with zero series resistance. For a given photocurrent, I_p, the equations governing the solar cell are

$$I = I_p - I_s \left[\exp\left(\frac{V}{V_o}\right) - 1 \right]$$

$$V_{oc} = V_o \ln\left(1 + \frac{I_p}{I_s}\right)$$

$$P_{out} = IV = I_p V - I_s V \left[\exp\left(\frac{V}{V_o}\right) - 1 \right] \tag{6.15}$$

where $V_o = kT/e = 25.8$ mV $(T = 300$ K$)$.

The condition for maximum power output is that a load resistance be connected which gives a cell voltage satisfying

$$\left(1 + \frac{V}{V_o}\right) \exp\left(\frac{V}{V_o}\right) = 1 + \frac{I_p}{I_s} \tag{6.16}$$

(see equation 6.6). Unfortunately, it is not possible to solve this equation in simple form. A simpler, but equivalent, procedure is to assume a particular cell voltage, and then to solve for the cell photocurrent which would give this voltage from equation 6.16. Substituting these values into equation 6.15, the maximum power obtainable from the solar cell for a given photocurrent may be calculated. Figure 6.14 is a plot of maximum power versus cell photocurrent for three different values of reverse bias saturation current, obtained in this way. The result is a nearly linear relationship between maximum power and cell photocurrent. This figure may be regarded as applicable to any solar cell with zero series resistance.

In order to get a feeling for just how much power may be extracted from a solar cell, consider an ideal cell which produces one electron for each incident photon. Such a cell would have unity quantum efficiency at all wavelengths.

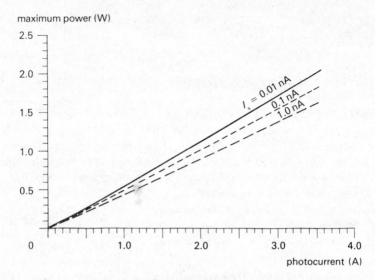

FIGURE 6.14

The variation of the maximum power obtainable from a solar cell (with zero series resistance) with photocurrent, I_p, for different values of reverse bias saturation current, I_s

The cell photocurrent would be

$$I_p = \int_0^\infty \frac{1}{E_0} P_{inc}(\lambda)\, d\lambda \qquad \left(\eta = 1, E_0 = \frac{hc}{\lambda e}\right)$$

For the extraterrestrial solar spectrum (AM0) the result is I_p = 2.524 A, assuming the cell area is 24.6 cm^2 as before. Using the same saturation current as for the solar cell of section 6.3 (I_s = 0.1 nA) the maximum power obtainable from the cell is P_{max} = 1.331 mW. The power conversion efficiency is η_{max} = 40 per cent.

This shows that even a solar cell which produces one electron for each incident photon at every wavelength does not produce 100 per cent power conversion efficiency. Why? One reason is that for a given material with bandgap energy E_g (E_g = 1.12 eV in silicon), many of the photons from the Sun have more energy than is required to release an electron for conduction. This additional energy is converted into the vibrational energy of the semiconductor lattice, but not into additional conduction electrons. In order to utilise all of the energy of the Sun's radiation, the ideal solar cell should produce more than one electron per photon at those wavelengths where the photon's energy exceeds the bandgap energy of the material.

Let us carry this argument a bit further. Suppose we have a semiconductor material called 'super silicon' which produces more than one electron from those

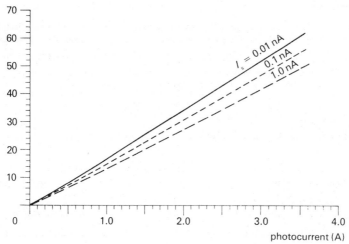

FIGURE 6.15

The variation of power conversion efficiency of the same solar cell as in figure 6.14 with photocurrent, based on the extraterrestrial solar constant (135.3 mW/cm^2); note that even the most idealised solar cell achieves a conversion efficiency of less than 50 per cent

photons with energy greater than the bandgap, E_g = 1.12 eV. Its quantum efficiency is

$$\eta = \frac{E_0}{E_g} \quad \text{(super silicon)} \tag{6.17}$$

The current responsivity of super silicon is $R_\phi = \eta/E_0 = 1/E_g = 0.893$ A/W, which is constant. A solar cell made of this material with a collection area of 24.6 cm^2 would yield a photocurrent given by $I_p = 2.972$ A under AM0 conditions. Again, using a saturation current $I_s = 0.1$ nA, the maximum power obtainable from the cell is $P_{max} = 1598$ mW. The power conversion efficiency is η_{max} = 48 per cent.

We see that even the most ideal solar cell, which could convert all of the energy of incident photons into electrical energy, produces a power conversion efficiency of only 50 per cent. The problem is that the solar cell is a forward biased $p–n$ junction. This has the effect of shunting some of the photocurrent, reducing the total cell current. But the most important effect is that the cell open circuit voltage is limited to the value at which the forward diode current equals the photocurrent. Since the forward diode current is proportional to the reverse bias saturation current, I_s plays a key role in determining the solar cell's power output. Unfortunately, once the semiconductor is doped to form a $p-n$ junction the reverse bias saturation current is fixed.

There seems to be little that can be done with currently available materials to overcome this fundamental limitation on solar cell power conversion efficiency. Efficiencies of 10 per cent are typical of the commercially available silicon solar cells. In the next section we shall consider the cost of producing electrical energy with solar cells.

6.9   The Economics of Solar Cells

As was mentioned in section 6.6, the amount of sunlight available for energy conversion is large even in regions far north, or south, of the equator. The longest daily average sunlight in summer occurs at the north pole, while the longest daily average in winter occurs at the south pole. Provided that local weather conditions do not create thick cloud cover for prolonged periods, northern latitudes can have a reasonably high yearly average sunlight.

A yearly average intensity of 200 W/m^2 could be expected in a country at a latitude of 50° N, such as the United Kingdom. A solar cell with an efficiency of 10 per cent could be expected to yield 20 W/m^2 averaged over 1 year, or a total energy output of 631 MJ/m^2 or 175 kW h/m^2. Thus, in the United Kingdom, depending on cloud conditions, a solar cell may be expected to yield as much as 175 kW h/m^2 in 1 year.

The current cost of domestic electricity in the United Kingdom is £27.18 per 1000 kW h. If solar cells are to compete commercially they should cost of the order of £4.76 per m^2 per year. Assuming a solar cell will last 10 years, a cost of £47.60 per m^2 would be competitive. Although the present cost of solar cells is approximately 26 times this figure, it is likely that their cost will fall rapidly as they come into widespread use.

With a home roof area of 20 m^2 covered with solar cells, a yearly average power equal to 400 W would be generated. This is sufficient for electric lighting and electrically heated hot water. Unfortunately, the domestic demand for electric lighting reaches a peak in the evening when little sunlight is available for conversion to electricity. This means that widespread domestic use of solar cells may require some method of storing electrical energy which is cheap and easily applied in the home. None the less, domestic electricity is used during daylight hours for domestic appliances, heating of hot water, and radio and television, and could be supplied in part by a large roof panel of solar cells.

The technology of electrical energy storage is still developing. The present method, by dry or wet cell battery, is expensive, bulky and not very efficient due to the rapid loss of charge of most batteries. A roof area of 20 m^2 and a yearly solar cell output of 631 MJ/m^2 gives 12 620 MJ in 1 year or 34.6 MJ per day. If this energy were stored in a capacitor at a potential of 20 V, it would have to have a capacitance of 172 900 F!

It seems likely that an electrical grid could distribute electrical power generated by solar cells to points of highest demand in communities. The problem of efficient storage of electrical energy remains to be solved by future generations of engineers.

Present applications of solar cells include the supply of electrical power in remote areas for microwave repeater stations, ocean navigational aids, communications, irrigation systems and other applications requiring a source of electrical power which is simple and easy to operate. Most of these applications consist of one or more solar panels charging a battery. Because of the solar cell current—voltage characteristic, the solar panel produces nearly maximum cell voltage over a wide range of solar intensity, and is therefore able to charge the battery throughout most of the day.

One very futuristic application of solar cells, which is being studied seriously in the United States and in Europe, is to put gigantic solar panels into orbit and to beam the collected power by microwave antenna to the surface of the Earth. An average intensity of 1353 W/m^2 exists at all times above the atmosphere, eliminating any need for energy storage. It is proposed that solar panels covering an area of 8 square miles be used to generate 5000 MW of electrical power, which would be beamed to Earth via 10 cm wavelength microwaves, for which the Earth's atmosphere is virtually transparent. If the microwave power is collected using a 5 mile diameter receiving antenna, a microwave power density of the order of 20 mW/cm^2 would be needed. It is estimated that sixteen such satellites in stationary orbit could provide the present electricity needs of the United Kingdom.

7 The Laser Diode

7.1 Introduction

The light emitting diode of chapter 2 can be modulated to act as the source in an optical communication system. The primary disadvantage of using an LED in such a system is that it emits light over a relatively wide range of wavelengths — the relative width of an LED's emission is typically around 5 per cent. One effect of this rather broad spectral emission is that because of dispersion in the optical transmission path the modulation bandwidth must be limited, so that different portions of the modulation spectrum do not arrive at different times at the receiver. If an LED which has a peak emission at 600 nm with a 5 per cent spectral width is used as the source in an optical communication system, the bandwidth of the source is 25 000 GHz. Trying to modulate such a source is rather like superimposing modulation on noise.

The idea of improving the light emitting diode by making it operate as a laser occurred in the early 1960s. The term laser is the acronym for 'light amplification by stimulated emission of radiation'. In 1962 several groups in the United States produced the first operational laser diodes, fabricated from a gallium arsenide (GaAs) $p–n$ junction diode. By fabricating the junction so as to have two parallel flat ends perpendicular to the plane of the junction, as in figure 7.1, the light emitted in the junction region was reflected back and forth, thus forming a standing wave pattern. When the junction is forward biased, conduction electrons from the n-layer cross into the conduction band of the p-layer. Within a short time these electrons recombine with holes in the p-layer, releasing photons with

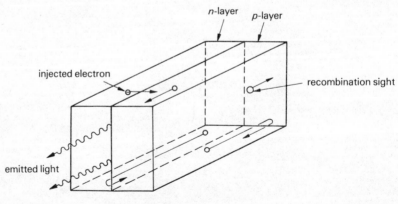

FIGURE 7.1
A simple $p–n$ junction laser

energy equal to the bandgap energy of the material. In GaAs the bandgap is 1.43 eV at room temperature and the wavelength of emitted photons is about 821 nm in the near infrared. Because of the build-up of photons reflecting back and forth between the two ends of the semiconductor crystal, the likelihood of a photon colliding with a conduction band electron is increased. When such a collision takes place stimulated emission may occur, resulting in the release of two photons in phase with one another. In this way an avalanche of photons builds up in the junction, all reflected back and forth between the two parallel flat ends.

The light emitted from the GaAs laser diode is coherent, it has a more or less uniform phase front and a relatively narrow spectral width of the order of about 1 nm. This represents an improvement of one or two orders of magnitude over the emission from an LED. The major problem with this simple $p-n$ junction laser is that the current density required to produce sufficient stimulated emission is so high that heating in the junction region will destroy the device unless it is operated in a pulsed way. A current density greater than 25 000 A/cm^2 is usually required, with pulse widths of less than 1 μs duration. The early laser diodes were cooled to the temperature of liquid nitrogen when operating, in order to overcome the heating problem.

7.2 Stimulated Emission

The process of stimulated emission was first postulated by Albert Einstein in 1917. The other two effects involving photons in a semiconductor are absorption and spontaneous emission (see figure 7.2). If a photon collides with an atom of

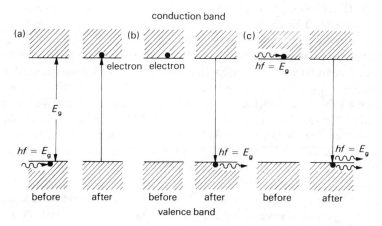

FIGURE 7.2

The three basic photon—electron transitions between valence and conduction bands in a semiconductor: (a) absorption; (b) spontaneous emission; (c) stimulated emission

the material it may be absorbed, causing one of the outer electrons to gain enough energy to pass into the conduction band. Absorption occurs most readily if the photon has energy equal to or greater than the bandgap of the material. Sometime later the freed electron will recombine with another atom, passing back into the valence band. When this happens a photon may be released, with energy equal to the amount the electron has just lost. This is called spontaneous emission. Stimulated emission occurs when a photon collides with a conduction electron. After the collision the electron can release two photons, exactly in phase with one another, and pass back into the valence band.

As was mentioned in chapter 1 the process of photon emission occurs most efficiently in a material such as GaAs which has a direct bandgap, so that no momentum is gained or lost in the transition. In silicon or germanium some energy and momentum may be transferred into vibrations of the crystal lattice, and although photon absorption occurs efficiently in these materials, photon emission does not.

In the GaAs laser diode the rate of stimulated emission must exceed the rate of photon absorption in order for lasing action to occur. This can only happen if there are more electrons in the conduction band than in the valence band of the material. This is called a population inversion. A population inversion can be created by injecting sufficient forward current through the junction, and usually more than 25 000 A/cm^2 are required. Thus a GaAs laser diode is operated with as much as several amperes of peak forward current.

7.3   The Double Heterostructure Laser Diode

The primary disadvantage of the laser diode as described above is the large current needed to achieve the population inversion necessary for laser action. This problem can be overcome in two ways; first, by confining the current to a smaller volume in the $p-n$ junction, thus increasing the current density for the same external current; second, by confining the light to a smaller volume, thus increasing the probability of stimulated emission for the same total power. It was a remarkable feat of engineering to find the solution to this problem which achieved both effects by a single modification of the laser diode. In the double heterostructure laser diode, a central layer of p-type GaAs is surrounded by layers of n-type and p-type AlGaAs, so as to form a sandwich which tends to confine the photons through total internal reflection, and which confines the electrons because the wider bandgap of AlGaAs acts as a potential barrier to reflect electrons. The semiconductor $Al_x Ga_{1-x} As$ (aluminium gallium arsenide) is formed by replacing some of the gallium atoms by aluminium atoms. Both aluminium and gallium have the same valence. If the proportion of gallium atoms which has been replaced is x, then the proportion of gallium atoms which remains is $1 - x$.

The bandgap energy of AlAs is 2.13 eV while that of GaAs is 1.43 eV.

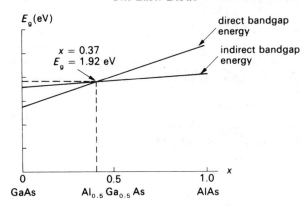

FIGURE 7.3

The direct and indirect energy gaps of $Al_x Ga_{1-x} As$

However, AlAs is an indirect bandgap material. In forming $Al_x Ga_{1-x} As$ the direct bandgap energy rises to 1.92 eV at $x = 0.37$ (see figure 7.3). For x greater than this value $Al_x Ga_{1-x} As$ is an indirect bandgap material. Photons with an energy characteristic of the bandgap energy of GaAs are less likely to be absorbed in $Al_x Ga_{1-x} As$ since it has a higher bandgap. For this reason the refractive index of $Al_x Ga_{1-x} As$ is smaller than that of GaAs.

In the double heterostructure laser diode, as shown in figure 7.4, alternate layers of GaAs and $Al_x Ga_{1-x} As$ are used to form a planar waveguide which confines the light through total internal reflection, since the refractive index of the central layer is greater than that of the layers on either side. The energy band structure is shown in figure 7.5. The change in bandgaps between the p-type gallium arsenide and p-type aluminium gallium arsenide acts as a potential barrier for electrons, confining them to the central layer, while the change in bandgaps between the n-type aluminium gallium arsenide and p-type gallium arsenide prevents holes crossing into the n-layers.

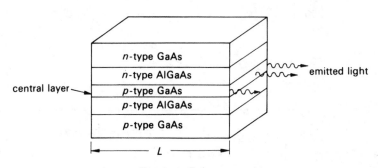

FIGURE 7.4

The double heterostructure laser diode

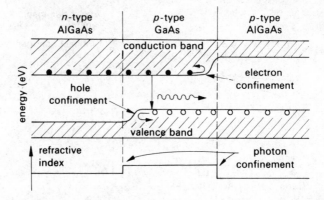

FIGURE 7.5

The bandgap diagram and refractive index profile of the AlGaAs–GaAs–AlGaAs layers

The minimum current density for laser action in the double heterostructure laser diode is about five times smaller than that needed in the simple two-layer gallium arsenide laser diode. The variation of threshold current needed for the onset of laser action with the thickness of the central GaAs layer is shown in figure 7.6. In this figure comparison is made between the double heterostructure laser and the single heterostructure laser, consisting of three layers of *n*-type gallium arsenide, *p*-type gallium arsenide, and *p*-type aluminium gallium arsenide.

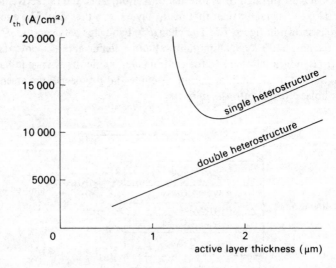

FIGURE 7.6

Variation of the threshold current density with the width of the active lasing region

Even though the current density required has been greatly reduced in the double heterostructure laser diode, the device must be bonded to an efficient heat sink if operated continuously at room temperature.

7.4 Radiation from a Stripe Geometry Double Heterostructure Laser Diode

Although the laser diode described above achieves efficient confinement and low current threshold, the carrier and optical confinement is in one direction only. In practice the device may emit light across a broad portion of the central gallium arsenide layer, resulting in a rather diffuse and irregular pattern of radiated light. This problem is overcome by confining the active region (the central gallium arsenide layer) to a narrow stripe, as shown in figure 7.7. Usually the active region is about 1 μm thick with a 10 μm (stripe) width. The resulting pattern of radiated light forms a uniform gaussian pattern as shown in figure 7.8. The near field pattern is the variation in light intensity at the front surface of the laser. It forms a gaussian spot of the form

$$P(x, y) = P_0 \exp[-(2x^2/x_0^2) - (2y^2/y_0^2)] \tag{7.1}$$

where P_0 is the intensity at the centre of the spot and x_0 is the beam radius in the x-direction, and y_0 the beam radius in the y-direction.

The total power emitted by the laser is the integral of the near field intensity over its front surface

$$P_T = \int_{-\infty}^{\infty} \int_{-\infty}^{\infty} P(x, y) \, dx \, dy \tag{7.2}$$

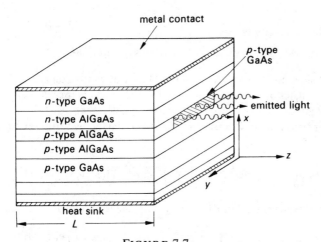

FIGURE 7.7

The stripe geometry double heterostructure laser diode

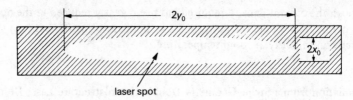

FIGURE 7.8

The near field intensity at the front surface of the laser diode

It may be easily shown using equation 7.1 that the total power is

$$P_T = \tfrac{1}{2}P_0(\pi x_0 y_0) \quad \text{(watts)} \tag{7.3}$$

In this expression $\tfrac{1}{2}P_0$ is the average intensity (with units of W/m^2) and $\pi x_0 y_0$ is the effective area of the gaussian spot.

The far field radiation pattern is the variation of emitted light intensity with angle, far from the front surface of the laser, as in figure 7.9. The far field intensity may be calculated from the two-dimensional spatial Fourier transform of the field amplitude at the front surface of the laser. Thus

$$E_0(K_x, K_y) = \int_{-\infty}^{\infty} \int \sqrt{P(x, y)} \exp[j(K_x x + K_y y)] \, dx \, dy$$

spatial	near
Fourier	field
transform	amplitude

$$= \sqrt{P_0(x_0 y_0 \pi)} \exp[-(K_x x_0/2)^2 - (K_y y_0/2)^2] \tag{7.4}$$

The far field intensity is

$$P_{FF}(\theta) = |E_0(K_x, K_y)|^2 \cos^2 \theta/(\lambda_0 r)^2 \tag{7.5}$$

($K_x = K_0 \sin\theta \cos\phi$, $K_y = K_0 \sin\theta \sin\phi$). In this expression r, θ and ϕ are the radius and angles of a spherical coordinate system, λ_0 is the wavelength of the emitted light, and $K_0 = 2\pi/\lambda_0$ is the wavenumber.

The far field intensity in the $y-z$ plane is obtained by setting $\phi = \pi/2$ in equation 7.5

$$P_{FF}(\theta) = P_0 \cos^2 \theta \left(\frac{\pi x_0 y_0}{\lambda_0 r}\right)^2 \exp\left\{-2[(K_0 y_0/2) \sin\theta]^2\right\} \tag{7.6}$$

($y-z$ plane) while the intensity in the $x-z$ plane is obtained by setting $\phi = 0$

$$P_{FF}(\theta) = P_0 \cos^2 \theta \left(\frac{\pi x_0 y_0}{\lambda_0 r}\right)^2 \exp\left\{-2[(k_0 x_0/2) \sin\theta]^2\right\} \tag{7.7}$$

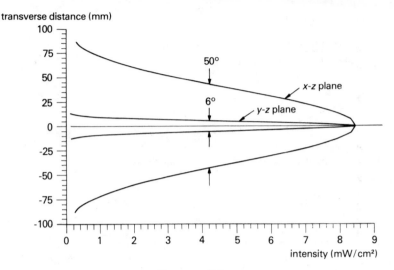

FIGURE 7.9

The horizontal and vertical radiation patterns of the stripe geometry double heterostructure laser

The radiation pattern of the double heterostructure laser diode is obtained by plotting profiles of constant intensity, as in figure 7.9. For example, in the $y-z$ plane the intensity is equal to 1 per cent of the intensity at the centre of the gaussian spot, along a profile defined by

$$r(\theta) = 10 \,\frac{\pi x_0 y_0}{\lambda_0}\, \cos\theta \,\exp\left\{-[(\pi y_0/\lambda_0)\sin\theta]^2\right\}$$
$$\text{(constant intensity profile in } y-z \text{ plane)} \quad (7.8)$$

The angular width of the radiation pattern is the angle $\theta_{1/2}$ where the intensity drops to half its maximum value on axis ($\theta = 0$). From equations 7.6 and 7.7 the angular width of the radiation pattern in the $y-z$ and $x-z$ planes is defined by

$$\cos\theta_{1/2}\,\exp\left\{-[(\pi y_0/\lambda_0)\sin\theta_{1/2}]^2\right\} = 1/\sqrt{2} \quad (y-z \text{ plane}) \quad (7.9)$$
$$\cos\theta_{1/2}\,\exp\left\{-[(\pi x_0/\lambda_0)\sin\theta_{1/2}]^2\right\} = 1/\sqrt{2} \quad (x-z \text{ plane}) \quad (7.10)$$

Note that at the angle $\theta = \theta_{1/2}$, the radius of the constant intensity profile is equal to $1/\sqrt{2}$ its value on axis ($\theta = 0$)

$$r(\theta_{1/2}) = 10 \,\frac{\pi x_0 y_0}{\lambda_0}\, \frac{1}{\sqrt{2}} = r(0)/\sqrt{2} \quad (7.11)$$

This result may be used to define the angular width of the radiation patterns of figure 7.10.

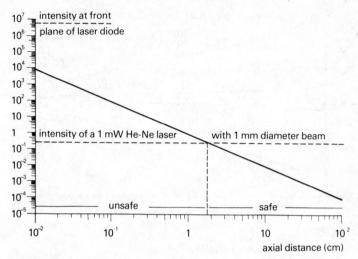

FIGURE 7.10
Variation of axial intensity of the laser emission with distance

The laser diode illustrated in this figure has an angular beamwidth of 6° on the horizontal $(y-z)$ and 50° in the vertical $(x-z)$ planes.

The beam radii which give these angular widths are $x_0 = 0.34$ μm and $y_0 = 2.96$ μm, at a wavelength $\lambda_0 = 850$ nm. If the laser emits a total power of 100 mW, the intensity at the centre of the gaussian spot is $P_0 = 6.33$ MW/cm^2. For comparison, the intensity of sunlight reaching the Earth's surface at the equator through dry clear air at noon on a summer day is about 100 mW/cm^2. Thus the laser's gaussian spot has a higher intensity, all concentrated at 850 nm wavelength.

At a distance $r = 1$ cm from the front surface of the laser, the intensity on axis is equal to $P_{FF}(\theta = 0, r = 1$ cm$) = 875$ mW/cm^2 while at $r = 10$ cm the intensity drops to 8.75 mW/cm^2 (see figure 7.10). We see in this simple example that even though the total emitted laser power is well above the level of laser eye damage, at a distance equal to 1 cm from the front surface of the laser the intensity has dropped to a safe level.

The laser emission may be collimated using a lens of large numerical aperture, as in figure 7.11. For example, a 2 cm diameter lens with 1 cm focal length will produce a collimated beam with diameters $W_s = 0.94$ cm and $W_T = 0.105$ cm. The power density at the centre of the collimated beam is 875 mW/cm^2, assuming the same example as above.

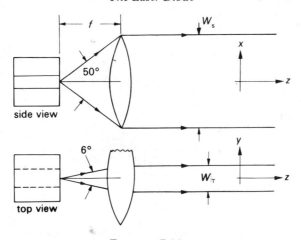

FIGURE 7.11
Collimation of light emitted from a laser diode

7.5 Spectral Properties of the Laser Diode

The central layer of the double heterostructure laser acts as a waveguide for the
light emitted through stimulated emission, as shown in figure 7.5. The light is
trapped through total internal reflection in the central GaAs layer, which has a
refractive index larger than that of the surrounding AlGaAs layers. The trapped
light is reflected back and forth between the opposite faces of the laser, a
distance L apart. The device forms an optical resonant cavity, producing an
emission spectrum with very narrow peaks as in figure 7.12. The basic condition
for resonance is that the change in phase of a photon which travels from one side
of the laser cavity to the other and back again, a total distance of $2L$, must be
an integer multiple of 360°. If the refractive index of the GaAs central layer is n,
then the speed of photons in this layer is c/n, where $c = 2.99 \times 10^8$ m/s in a
vacuum. The time taken to travel a distance $2L$ is $2L/(c/n)$ or $2nL/c$. If the
photon has frequency f then the phase change in travelling a distance $2L$ is

$$(2\pi f)\ \frac{2nL}{c} = 2m\pi \qquad \text{(integer multiple of 360°)} \qquad (7.12)$$

where m is an integer. Since each value of m gives a different resonant frequency,
it is usual to write

$$f = f_m = \frac{mc}{2nL} \qquad (m = \text{integer}) \qquad (7.13)$$

In practice resonances may occur for several different values of m since
photon emission occurs over a band of wavelengths centered around the bandgap

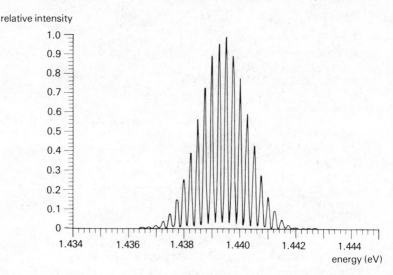

FIGURE 7.12

Spectral emission of a laser diode with many longitudinal modes

energy of the material. The relationship between the resonant frequency f_m and the wavelength of this resonance is $f_m = c/\lambda_m$. Each resonance is usually referred to as a longitudinal mode. Although it is possible to design a laser which operates with a single longitudinal mode, or a single resonant frequency, in practice most lasers have many of these modes together. The frequency interval between longitudinal modes is $\Delta f = f_{m+1} - f_m = c/2nL$.

In the laser diode of figure 7.12 a total of twenty longitudinal modes occurs between 1.437 eV and 1.442 eV. This gives a frequency interval between them of $\Delta f = 60$ GHz. The central line in the spectrum is at energy of 1.4394 eV or a frequency of approximately 350 THz (1 THz = 10^{12} Hz).

Each line in the spectrum of figure 7.12 represents a relatively pure resonant frequency of the laser diode. The width of these lines is typically less than 0.1 nm (1 Å), a value for the line width of 0.1 Å being perhaps most typical. In terms of frequency the line widths are of the order of $\delta f = 40$ GHz or less*
The spectral purity of one of the longitudinal modes of the laser diode is therefore better than one part in 10^4. In contrast to this the spectral purity of an LED is about one part in twenty. The spectral purity of a He–Ne gas laser is about one part in 10^8, and the spectral purity of an ordinary laboratory standard signal generator is about one part in 10^8. The spectral purity of the natural

---

* From the relationship between frequency and wavelength, $f = c/\lambda$, we obtain $\delta f/\delta \lambda = -c/\lambda^2$ or $\delta f/f = -\delta \lambda/\lambda$. If $\delta \lambda = 1$ A, $\lambda = 8570$ A, $f = 350$ THz, $\delta f = 41$ GHz.

emission from cadmium at a wavelength of 650 nm in the red is about one part in 10^6.

Although a spectral purity of one part in 10^4 is good, the emission spectrum of each of the longitudinal modes of the laser diode is a few tens of gigahertz wide! If the laser is used as a source in an optical communication system, the modulation signal is likely to have a smaller bandwidth than the linewidth of each of the laser lines. What is the physical meaning of a source such as the laser diode with such a large linewidth?

Figure 7.12 is a plot of the relative intensity of the laser emission against energy hf/e. Another interpretation of this plot is that it represents the probability density of emission of photons with a particular energy. This is illustrated in figure 7.13, which is a plot of the spectrum of a single longitudinal mode, normalised so that the total area under the curve is 1. The probability of emission of a photon with energy between E and $E + dE$ is $p(E)\, dE$, and the probability of emission of a photon with energy in the interval E_1 to E_2 is $P\{E_1 \leqslant E \leqslant E_2\} = \int_{E_1}^{E_2} p(E)\, dE$.

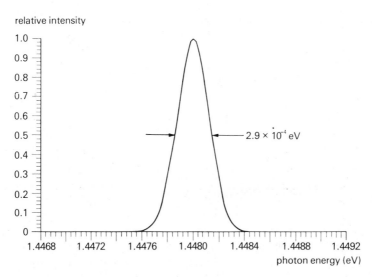

FIGURE 7.13
Probability density of the emission of photons of a single longitudinal mode of a laser diode

Since each photon is a wave packet with frequency f, the laser emission spectrum can also be interpreted as representing the probability density of emission of light at a particular frequency. We imagine that a particular longitudinal mode consists of a very large number of photons emitted per second, with frequencies distributed as in figure 7.14. These photons all combine to form an output light intensity which is a more or less continuous sinusoidal variation at frequency f_0. Because of internal changes in the active region of the laser diode, temperature fluctuations cause variations in refractive index and bandgap energy, the frequency and the amplitude of the emitted light all vary slightly in time. If the probability density of figure 7.14 is a normal distribution with half bandwidth $\delta f = 35$ GHz, centered at $f_0 = 350$ THz, then in a particular sample of the emitted light which is 10 000 cycles long, approximately 6000 cycles will have frequency between $f_0 - \delta f$ and $f_0 + \delta f$.

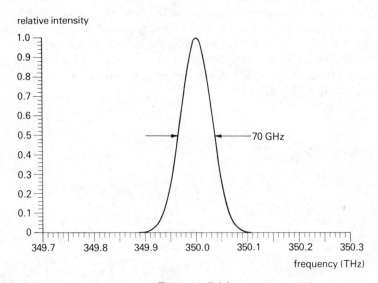

FIGURE 7.14

Probability density of photon emission at frequency f

Another way of expressing this is to say that the light emitted in a single longitudinal mode of the laser diode is correlated over about 10 000 cycles of the waveform. That is, if we compare different portions of the time variation of the emitted light, we find that different cycles of the sinusoidal variation are in phase up to about 10 000 cycles separation, but for a wider separation they are not. The correlation time — the maximum separation over which successive portions of the signal are in phase — is 10 000 cycles divided by 350 THz, or about 29 ps (1 ps = 10^{-12} s). Thus we find that a single longitudinal mode of a laser diode represents a sinusiodally varying waveform whose frequency drifts slightly because of internal variations of the device, and whose spectral purity of better than one part in 10^4 gives a correlation time of greater than 30 ps.

The above qualitative description can be made more precise by finding the

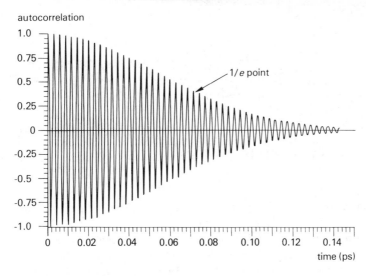

FIGURE 7.15

The autocorrelation of the laser emission spectrum of figure 7.14

inverse Fourier transform of the spectrum of figure 7.14. This is known as the autocorrelation of the longitudinal mode waveform. In this case the auto-correlation is a sinusoidally varying waveform with frequency f_0 whose envelope slowly decays as in figure 7.15. The time constant of the decaying envelope is the correlation time of the laser emission.

A He–Ne laser that emits light with a spectral purity of about one part in 10^8 has a correlation time of about 20 ns (1 ns = 10^{-9}s). A signal generator with a spectral purity of one part in 10^8 whose output is centered at 1 kHz, has a correlation time of 10^5 s or about 28 hours!

7.6 Coherence Length of the Laser Diode

The light emitted by the laser diode represents a sinusoidally varying electro-magnetic field which travels at approximately 3 x 10^8 m/s through air (see figure 7.16). The temporal variation discussed above, which results from the linewidth of the laser spectrum, is exactly mirrored in a sinusoidal spatial variation. It is quite accurate to think of the laser emission as representing a sinusoidal spatial variation, with wavelength $\lambda_m = c/f_m$, which travels along at the speed of light. For this reason the linewidth of a longitudinal mode determines a coherence length of the laser emission, which is just the speed of light times the correlation time. In the case of the laser diode discussed above

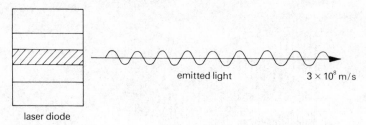

emitted light 3×10^8 m/s

laser diode

FIGURE 7.16
The spatial sinusoidal variation of the light emitted from a laser diode

the wavelength was 857 nm (f_0 = 350 THz), and a spectral purity of one part in 10^4 gives a coherence length of 9 mm.

The physical interpretation of the coherence length is shown in figure 7.17. The emitted light is split into two beams which travel different distances before being recombined at the photodetector. Since the output voltage from the photodetector is proportional to the intensity of light falling on it, which in this case is the square of the sum of the two beam amplitudes, the output voltage is proportional to the product of the amplitudes of two beams of light which have travelled a different distance. By varying the position of the second mirror, a plot of photodetector output voltage against distance gives the time average of the autocorrelation shown in figure 7.15, provided the relationship between distance and time is recognised.

In the case of the laser diode we should find that the photodetector output voltage would be reduced for path length differences greater than about 9 mm.

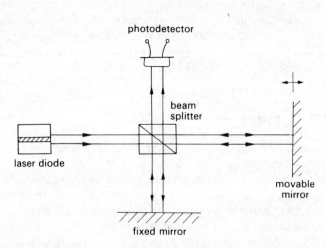

FIGURE 7.17
Experimental determination of the coherence length of a laser

7.7   The Effect of Many Longitudinal Modes

Normally a laser emits a large number of longitudinal modes. The effect of this is to produce a beat frequency equal to the line separation, which in figure 7.12 is about 60 GHz, which modulates the amplitude of the emitted light. For this reason the autocorrelation of the longitudinal mode spectrum is modulated as shown in figure 7.18, reducing the correlation time (or the coherence length) of the laser emission. If the beat frequency is 60 GHz the correlation time is reduced to less than 17 ps, and the coherence length is reduced to less than 5 mm.

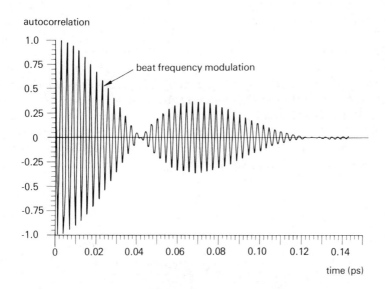

FIGURE 7.18

The autocorrelation of the emission spectrum of figure 7.12, showing the beating between different longitudinal modes, which reduces the coherence length of the laser

In practice the coherence length of the laser is limited by this effect. For example, a commercially available double heterostructure laser diode has a linewidth of about 4 nm centered at 850 nm. The cavity length is 100 μm, giving a longitudinal mode spacing of 417 GHz. A total spectral width of 4 nm implies a frequency width equal to 1600 GHz, implying that approximately four longitudinal modes are present in this laser. The coherence length of 417 GHz mode separation is about 0.72 mm.

7.8 Longitudinal Modes and Dispersion

Although the emission spectrum of a laser diode is relatively narrow, consisting of several longitudinal modes each representing a very high Q resonance of the optical cavity, the overall spectral width is much broader than any modulation spectrum that is likely to be transmitted in an optical communication system. It is likely that such a system will use pulse code modulation (PCM). The output from the laser diode will consists of a series of pulses of light, representing a digital coding of the modulation signal to be transmitted. In PCM a pulse represents the binary digit 1, while the absence of a pulse represents a 0. In the Post Office experimental optical communication link between Martlesham and Kesgrave in the United Kingdom, the pulses are transmitted at a rate of 140 Mbit/s over a 6 km length of optical fibre. Telephone conversations are coded and superimposed through multiplexing, and transmitted through the optical fibre.

The spectrum of the pulsed laser emission is the convolution of spectrum of the laser diode, for example as in figure 7.12, and the spectrum of the pulses from the PCM electronics. The pulses from the PCM electronics are more or less gaussian in shape (see figure 7.19), with a pulse width of around 1 ns (10^{-9} s). Consequently the bandwidth of the modulation pulses is about 1 GHz, much narrower than the linewidth of any of the longitudinal modes of the laser spectrum. When the relatively narrow spectrum of the PCM pulses is convolved with the laser spectrum, the laser spectrum is more or less unchanged. The pulses

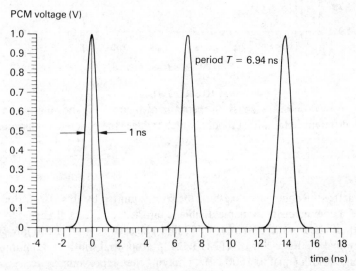

FIGURE 7.19
Modulation voltage from PCM electronics, with 144 MHz pulse rate

of light which are transmitted through the optical communication fibre have the same spectrum as that of figure 7.12, consisting of about twenty longitudinal modes spaced 60 GHz apart.

Each pulse travels in the optical fibre as a wave packet with a group velocity given by

$$V_g(\lambda) = \frac{c}{n(\lambda) - \lambda dn/d\lambda} \qquad (7.14)$$

where $c = 2.998 \times 10^8$ m/s and $n(\lambda)$ is the refractive index of the glass fibre at wavelength λ. Because glass is dispersive its refractive index varies with wavelength. Since each longitudinal mode is centered at a different wavelength, $\lambda_m = c/f_m$, each mode will travel with its own group velocity along the optical fibre. These velocities are given by

$$V_g(\lambda_m) = \frac{c}{n(\lambda_m) - \lambda_m d_n/d\lambda_m} \qquad (7.15)$$

At the photodetector, which may be a kilometre or more from the laser source, a series of pulses arrives at slightly different times for each pulse transmitted. If the distance between transmitter and receiver is d then the times of arrival of a pulse transmitted at $t = 0$ are $t_m = d/V_g(\lambda_m)$, for each longitudinal mode. This is illustrated in figure 7.20. The result of dispersion is that the transmitted pulses are broadened, setting an upper limit on the pulse rate, and hence the information capacity of an optical communication system.

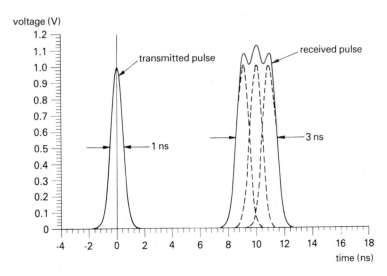

FIGURE 7.20
Pulse broadening due to dispersion in an optical communication system

The effect of dispersion in an optical communication system may be overcome in two ways. First, by using optical fibre materials and dimensions which minimise dispersion where it may be necessary to choose a particular wavelength (at which dispersion is minimum) for the laser diode. Secondly, by designing a laser which has a single narrow line; this has not yet been achieved in the commercial manufacture of the laser diode.

8 Optical Imaging Techniques

8.1 Introduction

The purpose of this chapter is to discuss optical imaging from the point of view of dynamic range, resolution and information capacity. Although photographic film has no proper place in a book on optoelectronic devices, we shall begin with a discussion of the properties of film in order to establish the basic principles of optical imaging. This is followed by a brief discussion of holography, an exciting new method of recording three-dimensional images. The remainder of the chapter deals with a new solid state imaging device, the charge coupled device (CCD) area image sensor. The resolution, dynamic range and information capacity of this device is considered in comparison with that of photographic film. The author concludes that although the density of sensing elements of the CCD area image sensor is low, its information capacity is reasonably high — because of the large dynamic range of silicon photodetectors — and that the problem is finding a method of encoding the optical image that can utilise this information capacity.

8.2 Photographic Film

Perhaps the best known optical imaging device is the photographic film. A typical high resolution photographic plate consists of a transparent glass base covered by a layer of photographic emulsion. The emulsion consists of a large number of photosensitive silver halide particles, uniformly distributed in a gelatin solution. When a silver halide particle absorbs sufficient light, it undergoes a chemical change which, after development of the film, allows it to be reduced to a tiny metallic silver particle, or grain; the grains which have not absorbed sufficient light remain unchanged, and are washed away in the development process, leaving a transparent region on the plate (see figure 8.1). The areas which received the most intense light have the highest concentration of silver grains, and look the darkest under illumination. The transmittance of the developed film therefore depends on the density of grains left after developing the film. In 1890 Hurter and Driffield showed that the area density of metallic silver grains of a developed film is related to the intensity transmittance of the film by

$$D = -\log_{10} T_i \tag{8.1}$$

where $T_i = I_0/I_i$ is the ratio of transmitted to incident intensity.

When a film is exposed to light the exposure of a given area of the emulsion is the product of the intensity of light falling on that area times the exposure

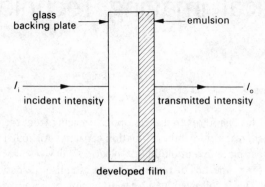

FIGURE 8.1
Transmission of light through a developed photographic plate

time, $E = It$. This is the total optical energy falling on a given area of the film during an exposure. The experimental relationship between the area density and the exposure is given by the Hurter–Driffield (H and D) curve (see figure 8.2). The slope γ of the linear portion of the H and D curve depends not only on the emulsion, but also on the developing process. High contrast film has a large value of γ.

In the linear portion of the H and D curve, the relationship between area density and exposure is

$$D = \gamma \log_{10} E - D_0 \tag{8.2}$$

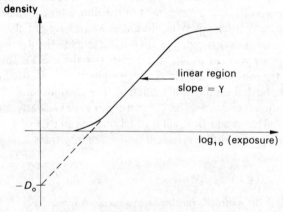

FIGURE 8.2
The Hurter–Driffield curve

Since $D = -\log_{10} T_i$, the intensity transmittance is related to the exposure by

$$T_i = K_o E^{-\gamma}, \qquad D_o = \log_{10} K_o \tag{8.3}$$

Thus the intensity transmittance is a nonlinear function of the exposure.

In many cases the amplitude transmittance is specified instead of the intensity transmittance

$$T_a = T_i^{1/2} = \text{amplitude transmittance} \tag{8.4}$$

The T–E curve, most often used in holography, is the experimental relationship between amplitude transmittance and exposure (see figure 8.3). In the linear portion of the H and D curve, this is

$$T_a = K_o^{1/2} E^{-\gamma/2} \tag{8.5}$$

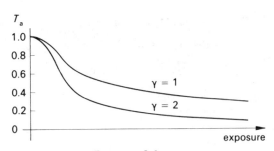

FIGURE 8.3

The variation of amplitude transmittance with exposure

8.3 Noise on Photographic Film

The photographic emulsion that has been exposed and developed consists of a large number of metallic silver grains, distributed in a thin layer over the surface of the backing plate. Assuming that the grain size is uniform and that the grains were uniformly distributed in the emulsion before exposure, the surface of the film is divided into cells representing the minimum resolvable image. Each cell contains M grains (see figure 8.4). Suppose that a given cell which has been exposed and developed has n developed grains and m undeveloped grains ($m + n = M$). There will be an unpredictable and random variation in the number of developed grains per cell even with uniform illumination, since the number of grains will vary from trial to trial due to variations in the emulsion and other external influences. These random variations in the number of grains per cell constitute the noise of the photographic film. In order to quantify this noise we must consider the statistics of the numbers of developed and undeveloped grains per

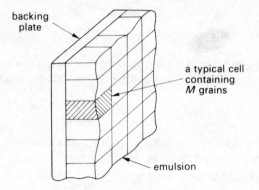

backing plate

a typical cell containing *M* grains

emulsion

FIGURE 8.4

The hypothetical division into cells of a photographic emulsion

cell, in particular their mean and standard deviation

$$\overline{m} = \text{the mean of } m$$

$$\sigma_m{}^2 = \overline{(m - m)^2} = \text{the variance of } m$$

$$\overline{n} = \text{the mean of } n$$

$$\sigma_n{}^2 = \overline{(n - n)^2} = \text{the variance of } n \tag{8.6}$$

Note that since $m + n = M$

$$\overline{n} = M - \overline{m}$$

$$\sigma_n{}^2 = \sigma_m{}^2 \tag{8.7}$$

A study of the characteristics of grain noise on photographic film* shows that the standard deviation in the number of developed grains, for a given mean number of developed grains, $\overline{n}$, is

$$\sigma_n = \sqrt{\left(\frac{\overline{m}\,\overline{n}}{M}\right)} = \sigma_m \tag{8.8}$$

The largest standard deviation, and hence the greatest noise, occurs at medium exposure (with $\overline{m} = \overline{n} = M/2$; $\sigma_n = \sqrt{M}/2$). Both at very low and very high exposure the noise is reduced. The ratio of standard deviation to mean gives the relative rms fluctuation in the number of developed grains per cell. At medium exposure

$$\frac{\sigma_n}{\overline{n}} = \frac{1}{\sqrt{M}} \quad \text{(medium exposure)} \tag{8.9}$$

* See F. T. S. Yu, *Introduction to Diffraction, Information Processing, and Holography*, (MIT Press, Cambridge, Mass., 1973).

This result is analogous to shot noise, where the relative fluctuation varies inversely with the square root of the number of charges generated per unit time.

A photographic film which is composed of a very large number of very small grains will have a small relative noise. For example, with $M = 10$, $\sigma_n/\bar{n} = 0.316 = 31.6$ per cent at medium exposure, while with $M = 100$ grains per cell, $\sigma_n/\bar{n} = 0.1 = 10$ per cent. Although the concept of a cell may seem rather arbitrary – for a given film we could define the cell to contain 10 or 100 grains – in practice other considerations will determine the minimum cell size. For instance, for a given wavelength of light a minimum distance between resolvable images may be defined, and this depends on diffraction effects which are independent of the properties of the film. Usually the resolution of the film is tested by photographing a series of more and more closely spaced lines. A resolution of 2000 lines/mm is considered good, while a low resolution film may record no more than 20 lines/mm.

8.4 The Signal-to-noise Ratio of Film

Although we have defined the noise of a film in terms of grain noise, it is more appropriate to consider the amplitude transmission of the developed film. The density of the developed film is related to the number of developed grains per cell, n. Referring to section 8.2, the amplitude transmittance is

$$T_a = 10^{-D/2} = 10^{-n/2} \tag{8.10}$$

Obviously T_a is subject to noise, since n varies randomly with exposure.

The statistics of the amplitude transmission may be defined by calculating the mean and standard deviation of T_a. Consider the Taylor expansion of T_a with n, about $n = \bar{n}$

$$T_a(n) = 10^{-n/2} = 10^{-\bar{n}/2} + \frac{d}{d\bar{n}} 10^{-\bar{n}/2}(n - \bar{n}) + \cdots$$

Using

$$\frac{d}{dn} 10^{-n/2} = -\tfrac{1}{2} \ln 10 \, 10^{-n/2} = -1.15 \, 10^{-n/2}$$

$$T_a(n) = 10^{-\bar{n}/2} - \tfrac{1}{2}\ln 10 \, 10^{-\bar{n}/2}(n - \bar{n}) + \cdots \tag{8.11}$$

The mean value of T_a since $\overline{n - \bar{n}} = \bar{n} - \bar{n} = 0$, is

$$\overline{T_a} = 10^{-\bar{n}/2} \tag{8.12}$$

Similarly $T_a - \overline{T_a} = -\tfrac{1}{2}\ln 10 \, 10^{-\bar{n}/2}(n - \bar{n})$ and the variance of T_a is

$$\overline{(T_a - \overline{T_a})^2} = (\tfrac{1}{2}\ln 10)^2 \, 10^{-\bar{n}} \, \overline{(n - \bar{n})^2}$$

Referring to equation 8.6 $\sigma_n^2 = \overline{(n - \bar{n})^2}$ and the standard deviation in T_a is

$$\sigma_{T_a} = \sqrt{[\overline{(T_a - \overline{T_a})^2}]} = \tfrac{1}{2}\ln 10 \, 10^{-\bar{n}/2} \sigma_n \tag{8.13}$$

If we regard the photographic signal in a given cell as the mean of T_a with power density $\overline{T_a}^2$, and the noise as the fluctuation about the mean with power density $\sigma_{T_a}^2$, the signal-to-noise ratio of a cell of the film is

$$\frac{S}{N} = \frac{\overline{T_a}^2}{\sigma_{T_a}^2} = \frac{1}{(\frac{1}{2}\ln 10)^2 \sigma_n^2} \tag{8.14}$$

The signal-to-noise ratio varies from cell to cell over the film. The minimum values will occur in cells which have received medium exposure $(\bar{n} - M/2)$, where $\sigma_n^2 = M/4$. Thus

$$\left(\frac{S}{N}\right)_{\min} = \frac{4}{(\ln 10/2)^2 M} = \frac{3.02}{M} \qquad \text{(medium exposure)} \tag{8.15}$$

In general a better signal-to-noise ratio is obtained with as few grains per cell as possible, hence the tendency toward 'grainy' photographs among professional photographers.

8.5 Information Capacity of Photographic Film

We have found that when a given cell of a photographic plate is exposed, the resulting transmittance may be larger or smaller than expected. The mean and standard deviation of the amplitude transmittance are

$$\overline{T_a} = 10^{-\bar{n}/2}$$

$$\sigma_T = \frac{1}{2}\ln 10 \sqrt{\left(\frac{\overline{m}\,\overline{n}}{M}\right)} \; \overline{T_a} \tag{8.16}$$

where $\bar{n}$ = mean number of developed grains per cell, $\overline{m}$ = mean number of undeveloped grains per cell and $\overline{m} + \bar{n} = M$ = total number of grains per cell.

Since there is a definite uncertainty in the transmittance of a cell of the photographic plate, the variety of images which may be recorded is somewhat limited. The amount of information which may be stored on a photographic plate is expressed by its information capacity, C.

Consider the information capacity of a rectangular area. If it has no distinguishing features its information content is zero. Suppose it is divided into two areas which may be either shaded or unshaded. The number of possible images which may be represented by these two areas is $2^2 = 4$, and the information capacity is $C = \log_2 2^2 = 2$ bits, as in figure 8.5.

The basic unit of information is the binary digit or bit. A single area which may be shaded or unshaded has an information capacity of 1 bit. Three areas which may be shaded or unshaded can represent eight images, with a capacity $C = 3$ bits (see figure 8.6).

If a rectangle is divided into $K \times L$ cells which may be either shaded or

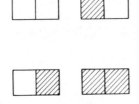

FIGURE 8.5

The four possible images of two areas which may be either shaded or unshaded

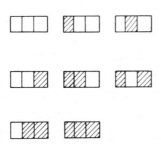

FIGURE 8.6

The eight possible images of three areas which may be shaded or unshaded

unshaded, the information capacity is

$$C = KL \text{ bits} \qquad \text{(2-level cells)}$$

$$KL = \text{total number of cells} \tag{8.17}$$

We must next consider the possibility that individual cells can have degrees of shading. The information capacity of a 3-level area, which may be unshaded, shaded, or cross shaded, is $C = \log_2 3 = 1.59$ bits. The information capacity of a pair of 3-level cells is $C = 2\log_2 3$, since there are nine possible images.

If each cell can have q levels of shading, the information capacity of KL cells is

$$C = KL \log_2 q \text{ bits} \qquad \text{(q-level cells)}$$

$$KL = \text{total number of cells} \tag{8.18}$$

A total of $2^C = q^{KL}$ different images may be recorded by a rectangular area with KL cells and q levels of shading.

In the case of photographic film the amplitude transmittance is quantised according to the number of grains per cell of the emulsion. Thus the shading is quantised into $M + 1$ different levels. However, as M increases so does the noise, represented by the variance of the number of developed grains per cell, σ_n^2.

The number of distinguishable levels is limited by the size of $\sigma_n{}^2$. Hence the information capacity of photographic film is limited.

Consider the density of a cell which has m grains

$$D = \log_{10} \overline{T_a{}^2} = \bar{n}$$

$$\sigma_n{}^2 = \frac{\overline{mn}}{M}$$

The largest noise occurs at medium exposure where $\bar{n} = M/2$. The maximum grain noise is $\sigma_n = \sqrt{M}/2$, and the effective number of levels of shading is

$$q = \frac{M}{\sigma_n} = 2\sqrt{M} \qquad \text{(number of levels for film)} \qquad (8.19)$$

If the total area of a rectangle is ab and the minimum cell size is d, then with q-levels of shading its information capacity is

$$C = \frac{ab}{d^2} \log_2 q \qquad (8.20)$$

In the case of a photographic emulsion which has N grains per unit volume, a cell of area d^2 and depth w has $M = Nwd^2$ grains. The number of grains per unit cell area is $M_0 = Nw$, thus $d^2 = M/M_0$.

Using the above expression for q, as in equation 8.19, the information capacity of a photographic plate of area ab is

$$C = ab \left(\frac{M_0}{M} \right) \log_2 (2\sqrt{M}) \qquad (8.21)$$

For example, suppose the maximum resolution of the film is 2000 lines/mm so that $M_0 = 2000^2$ grains/mm^2. Then

$$C = ab \left(\frac{4 \times 10^6}{M} \right) \log_2 (2\sqrt{M})$$

The maximum information capacity is obtained with $M = 1$ grain per cell, giving

$$C = ab \times 4 \times 10^6 \text{ bits}$$

With a plate area $a = 30$ mm x $b = 50$ mm, the information capacity of high resolution film is

$$C = 6 \times 10^9 = 6 \text{ Gbits} \qquad (M = 1 \text{ grain per cell})$$

If the number of grains per cell is $M = 16$, the information capacity is 1.125 Gbits, while with $M = 64$ grains per cell $C = 0.375$ Gbits.

This example shows that as the number of grains per cell increases, the information capacity of the film is reduced, consistent with the result that the signal-to-noise ratio of film is reduced as M increases. Both the largest signal-to-noise ratio and the greatest information capacity are achieved with 1 grain per cell on the film.

8.6   Holography

The image recorded by a photographic plate in traditional photography is a two-dimensional view of some scene. The enormous information capacity of the film, around 4 Mbits/mm^2 in high resolution film, is not utilised in most cases. In fact, a method of encoding optical images which makes use of the information capacity of film remains to be discovered.

Recently a method of recording three-dimensional images, which goes some way towards utilising the information capacity of film, has been developed by Dennis Gabor and others (see figure 8.7). In a hologram the emulsion records the interference pattern formed when a reference light source and light reflected from an object are combined. This is possible only if coherent light such as that produced by a laser is used. The reflected light from an object has an extremely complicated amplitude and phase variation across the photographic plate which, when combined with a reference beam, yields a complex interference pattern on the plate (see figure 8.8). After developing the film the resulting negative forms a swirling pattern of light and dark bands, which can diffract a laser beam.

The three-dimensional image of the original object may be viewed by re-illuminating the developed plate as shown in figure 8.9. The object can be seen in its original position by looking through the plate. Diffracted light as seen by the eye forms an image which is precisely like the light reflected from the original object, giving the impression of viewing the original in its full three-dimensionality.

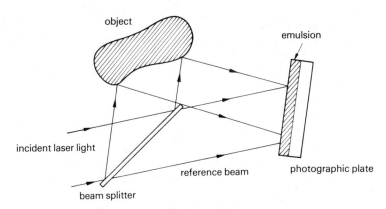

FIGURE 8.7

A method of recording the three-dimensional image of an object; the resulting developed photographic plate is called a hologram

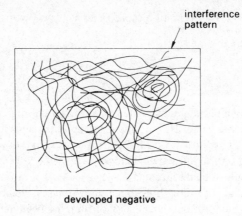

FIGURE 8.8
The interference pattern recorded on a hologram

A simple analysis of the interference pattern formed in a hologram may be obtained by considering two plane waves overlapping on a plate (the photographic emulsion) as shown in figure 8.10. Suppose the two waves are incident on the plate at angles $\pm\theta_0$ relative to the normal. Thus

$$E_1 = A_1 \cos\left[\omega t + k \sin\theta_0 x - k \cos\theta_0 z\right]$$
$$E_2 = A_2 \cos\left[\omega t - k \sin\theta_0 x - k \cos\theta_0 z\right] \qquad (8.22)$$

where $k = 2\pi/\lambda = \omega/c$, λ = the wavelength of both plane waves, and $c = 2.998 \times 10^8$ m/s, which is the speed of light in a vacuum. The intensity of

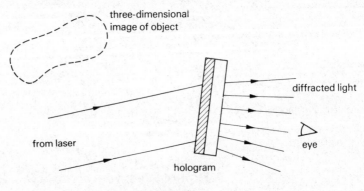

FIGURE 8.9
Viewing the hologram

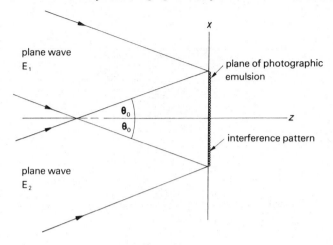

FIGURE 8.10
Recording the interference pattern of two plane waves

light on the plate at $z = 0$ is

$$I \sim |E_1 + E_2|^2 = |E_1|^2 + |E_2|^2 + 2E_1 E_2$$

The three terms are

$$|E_1|^2 = A_1^2 \cos^2(\omega t + k \sin \theta_0 x) = A_1^2 [1 + \cos(2\omega t + 2k \sin \theta_0 x)]/2$$

$$|E_2|^2 = A_2^2 \cos^2(\omega t - k \sin \theta_0 x) = A_2^2 [1 + \cos(2\omega t - 2k \sin \theta_0 x)]/2$$

$$2E_1 E_2 = 2A_1 A_2 \cos(\omega t + k \sin \theta_0 x) \cos(\omega t - k \sin \theta_0 x)$$

$$= A_1 A_2 \cos(2\omega t) + A_1 A_2 \cos(2k \sin \theta_0 x)$$

If the photographic plate is exposed over time T, the rapidly oscillating terms such as $\cos(2\omega t)$ average to zero. Thus the average intensity recorded on the plate is

$$I_{av} = \frac{A_1^2}{2} + \frac{A_2^2}{2} + A_1 A_2 \cos(2k \sin \theta_0 x)$$

Defining

$$I_o = \frac{A_1^2}{2} + \frac{A_2^2}{2}, \qquad m = \frac{2A_1 A_2}{(A_1^2 + A_2^2)}$$

$$I_{av} = I_o [1 + m \cos(2k \sin \theta_0 x)] \qquad (8.23)$$

Consider the variation of I_{av} over the surface of the plate. Using $k = 2\pi/\lambda$, $2k \sin \theta_0 = 4\pi \sin \theta_0/\lambda$. A series of light and dark bands is formed. The distance between successive dark bands is $\lambda_s = \lambda/2 \sin \theta_0$. The spatial frequency recorded

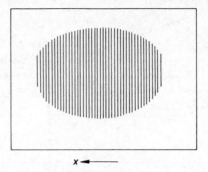

$x \longleftarrow$

FIGURE 8.11

Hologram showing the interference pattern of two plane waves

on the film, in lines per unit length, is

$$f = \lambda_s^{-1} = \frac{2 \sin \theta_0}{\lambda} \tag{8.24}$$

For instance, with $\theta_0 = 10°$ and $\lambda = 632.8$ nm, $\lambda_s = 1.82 \ \mu$m or $f = 549$ lines/mm.

The spacing between dark lines in the interference pattern must be large enough to be recorded on the film, which has its own resolution (see figures 8.11 and 8.12). The high resolution film used in holography can record of the order of 2000 lines/mm. The resolution of the photographic plate limits the field of view in recording a hologram. The maximum angle for 2000 lines/mm is $\theta_0 = 39°$.

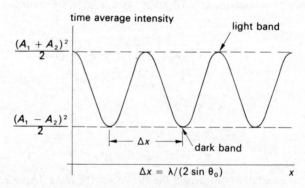

FIGURE 8.12

Variation of intensity with distance in the hologram of figure 8.11

8.7   Viewing the Hologram

Returning to the example of a hologram of a plane wave, the exposure, which determines the density of the negative, is the product of average intensity and exposure time T

$$E = I_{av}T = I_0 T \left[1 + m \cos(2k \sin \theta_0 x)\right] \tag{8.25}$$

The amplitude transmittance of the photographic plate, assuming the average exposure, $I_0 T$, falls in the linear region of the H and D curve, is

$$T_a = k_0^{1/2} E^{-\gamma/2} = T_a(E)$$

Substituting for the exposure from equation 8.25

$$T_a = 0.5 \left[1 + m \cos(2k \sin \theta_0 x)\right]^{-\gamma/2} \tag{8.26}$$

The value 0.5 indicates that the average exposure was at the centre of the $T-E$ curve $(k_0^{1/2}(I_0 T)^{-\gamma/2} = 0.5)$.

A study of equation 8.26 shows that the developed hologram forms a negative of the original exposure, since when the exposure is maximum $[E = I_0 T(1 + m)]$ the amplitude transmittance is minimum $[T_a = 0.5 (1 + m)^{-\gamma/2}]$, and vice versa. That is, light bands become dark bands and vice versa.

Another effect of the film is that it records a nonlinear impression of the exposure. Too much exposure causes clipping of the cosine modulation of the average intensity, which may give a modulation of the amplitude transmittance resembling a square wave. The additional harmonics introduced by nonlinear distortion of the exposure can cause spurious images when the hologram is viewed. If we think of the hologram as a diffraction grating, then when it is reilluminated there may occur a series of diffracted images in addition to the one recorded.

The amplitude transmittance of equation 8.26 is a periodic function with period $\Delta x = \lambda/2 \sin \theta_0$ which may be expanded as a Fourier series. The coefficients will depend on m and γ in a nonlinear fashion. Assuming the precise values of the Fourier coefficients may be determined, the amplitude transmittance may be written as

$$T_a = a_0(m, \gamma) + \sum_{n=1}^{\infty} a_n(m, \gamma) \cos(2nk \sin \theta_0 x) \tag{8.27}$$

where

$$a_0(m, \gamma) = \frac{1}{\Delta x} \int_0^{\Delta x} T_a(x) \, dx, \qquad \Delta x = \lambda/2 \sin \theta_0$$

$$a_n(m, \gamma) = \frac{1}{\Delta x} \int_0^{\Delta x} T_a(x) \cos(2nk \sin \theta_0 x) \, dx$$

The developed hologram may be viewed by reilluminating with a plane wave in the same direction as one of the original pair. Suppose the illumination is

$$E_0 = A_0 \cos(\omega t + k \sin \theta_0 x - k \cos \theta_0 z) \qquad (8.28)$$

Multiplying by the amplitude transmittance of equation 8.27 gives the amplitude of illumination on the holographic plate

$$E_0 T_a = a_0 A_0 \cos[\omega t + k \sin \theta_0 x] + \tfrac{1}{2} A_0 \sum_{n=1}^{\infty} a_n \cos[t - (2n-1)k \sin \theta_0 x]$$

$$+ \tfrac{1}{2} A_0 \sum_{n=1}^{\infty} a_n \cos[\omega t + (2n+1) k \sin \theta_0 x] \qquad (8.29)$$

Each term represents a diffraction pattern on the surface of the plate which propagates a plane wave in a particular direction. The first term, called the zero order diffraction image, is that portion of the illumination which passes straight through the photographic plate. Consider one of the other terms, say $\tfrac{1}{2}A_0 a_n$ $\cos[\omega t - (2n-1) k \sin \theta_0 x]$. Provided that $(2n-1) \sin \theta_0 \leqslant 1$ this term propagates a plane wave in front of the plate of the form

$$\tfrac{1}{2}A_0 a_n \cos[\omega t - (2n-1) k \sin \theta_0 x - K_n Z] \qquad (8.30)$$

where $K_n = \sqrt{[k^2 - (2n-1)^2 k^2 \sin^2 \theta_0]}$. This plane wave travels at an angle $\theta_n = \sin^{-1}[(2n-1) \sin \theta_0]$ to the z-axis. If $(2n-1) \sin \theta_0 > 1$ then the plane wave attenuates in the z-direction

$$\tfrac{1}{2}A_0 a_n \cos[\omega t - (2n-1) k \sin \theta_0 x] \exp(-X_n Z) \qquad (8.31)$$

where $X_n = \sqrt{[(2n-1)^2 K^2 \sin \theta_0 - k^2]}$. Any attenuating terms would not be visible except very near the surface of the plate.

We see that the illuminated hologram produces a collection of plane waves, most of which attenuate away from the surface of the plate. In this example the real image is that term which travels in the same direction as the original plane wave E_2, obtained with $n = 1$ in equation 8.30

$$\tfrac{1}{2}A_0 a_1 \cos[\omega t - k \sin \theta_0 x - k \cos \theta_0 z] \qquad \text{(real image)}$$

The other terms are spurious images, most of which will not be visible. This is illustrated in figure 8.13 in the case where $\theta_0 = 10°$.

A virtual image also appears behind the plate, due to light diffracted back (instead of forwards) from the hologram. The virtual image is quite real but must be viewed from behind the hologram. A series of reflected spurious images will also appear behind the hologram.

Spurious images are in general an undesirable feature of a hologram since they may overlap and blur the real image. This problem is minimised by using a reference intensity larger than the object intensity so that $m \ll 1$. The spurious images will be proportional to the square or higher powers of m, and therefore

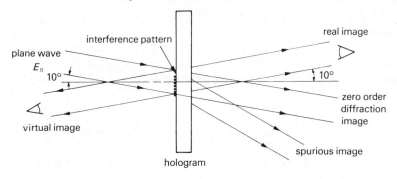

FIGURE 8.13

Viewing the hologram of a plane wave with $\theta_0 = 10°$

will have negligible amplitude. Using a Taylor expansion of the amplitude transmittance assuming $m \ll 1$, the illumination of the hologram is

$$E_0 T_a = \underbrace{0.5A_0 \cos(\omega t + k \sin \theta_0 x)}_{\text{transmitted illumination}} - \underbrace{0.125A_0 \gamma m \cos(\omega t - k \sin \theta_0 x)}_{\text{real image}}$$

$(m \ll 1)$

$$\underbrace{- 0.125A_0 \gamma m \cos(\omega t + 3k \sin \theta_0 x)}_{\text{spurious image}}$$

In this case the hologram produces one spurious image.

The enormously detailed diffraction pattern of a hologram goes some way towards utilising the large information capacity of a photographic plate. This utilisation may be increased even further by recording a series of holograms on the same plate at different angles of the reference beam. These different images may be viewed by either reilluminating at different angles, or by viewing the plate from different angles. A moving picture may be recorded in this way.

8.8   The Modulation Transfer Function

So far we have spoken of the resolution of an imaging device as if it has one value in all situations, for instance 2000 lines/mm for film. In practice what happens is that the finer detail of an image appears fainter than in the original. If we think of a two-dimensional image as having a spectrum, in the same way that a signal has a Fourier spectrum, then it is reasonable to consider the effect of the imaging device on a single spatial frequency. Suppose that the imaging device is to record the intensity

$$I_{av} = I_0\left(1 + m \cos\left[\frac{2\pi x}{\lambda_s}\right]\right) \qquad \text{(recorded image)} \qquad (8.32)$$

which would result from the interference pattern of two plane waves.

When this intensity is recorded the resulting image may be projected and analysed. As we have already seen in the case of photographic film, the imaging device may have nonlinear properties which will cause additional spurious images. Assuming these undesirable spurious images may be eliminated by some means, the projected image, which represents the original sinusoidally modulated intensity, will take the form

$$I' = I_0' \left(l + m' \cos \left[\frac{2\pi x}{\lambda_s} \right] \right) \qquad \text{(projected image)} \qquad (8.33)$$

Generally both the average intensity and the degree of modulation are affected by the imaging device.

The change between the projected and recorded images may be quantified by defining the modulation transfer function (MTF)

$$\text{MTF} = \frac{m' \ \text{(projected image)}}{m \ \text{(recorded image)}} \qquad (8.34)$$

In the case of photographic film the series of parallel lines represented by the recorded image will be somewhat scrambled by the granularity of the film. The variation of the modulation transfer function with spatial frequency λ_s^{-1} is shown in figure 8.14 for high resolution film. We see that the finer detail, which would have a high spatial frequency, will be fainter in the projected image.

8.9   The Charge Coupled Device Area Image Sensor

The idea of a solid state imaging device consisting of an array of photodetectors (photodiodes or phototransistors) on a silicon chip is extremely appealing. With

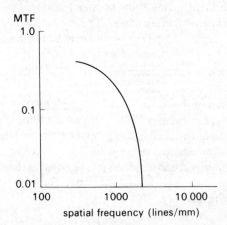

FIGURE 8.14

The variation of modulation transfer function of high resolution film with spatial frequency

the miniaturisation which can be achieved with large scale integrated circuits it might be thought that an extremely compact imaging device could be fabricated in this way. If each element of the array of photodetectors was 10 μm x 10 μm, then a 600 x 600 element array, which would be more than adequate for television, would require a total area of 6 mm x 6 mm. This imaging area could act as the sensor of a miniature camera, whose output could be recorded on video tape. A problem with this is that each element of the photodetector array would have to be connected to the appropriate electronics so that it could be scanned line by line. The compactness of the interconnections, which leads to problems of shunt capacitance and signal cross-talk, make this approach rather unsatisfactory.

The principle of charge transfer from element to element provides a practical solution to constructing a workable solid state imaging device. This is illustrated in figure 8.15. A series of closely spaced metal-oxide-semiconductor (MOS) capacitors is fabricated with every third element connected to the same common terminal. If a given element is properly biased it can store charge (holes or electrons). By an appropriate sequence of voltage pulses applied to each of the three terminals, the stored charge is transferred from one element to the next.

The stored charge could be injected by some external means. In fact the CCD was invented in order to provide a semiconductor rival to the magnetic bubble computer memory. It was thought that the presence or absence of charge could represent the binary digits 1 and 0.

The charge coupled device may be used for optical imaging because incident photons absorbed in the depletion region of one of the biased elements generate additional charge. The amount of charge stored is proportional to the intensity of light incident on the element during the storage time. A large number of closely spaced columns of elements form an imaging area.

It is found that individual columns of the array can be read out very rapidly, since the charge of each element can be rapidly transferred. Also, the efficiency

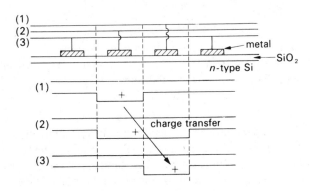

FIGURE 8.15
The principle of operation of the charge coupled device

of transfer of charge from element to element is high, of the order of 99 per cent or higher. Obviously the rate of transfer and the charge transfer efficiency are related, since at very high speed there may not be sufficient time for an element to completely discharge.

Several manufacturers have produced CCD imaging devices for use in a video camera. Typical of these is the Fairchild 190 x 244 element area image sensor. This consists of 190 columns of 244 elements each interlaced with storage registers. The stored charge is transferred column by column to a horizontal output register and then through a video preamplifier to a video output terminal. The video preamplifier and imaging array are fabricated on a single silicon chip so as to make a compact unit which produces around 200 mV of signal at its output. The image sensing elements measure 14 μm horizontally by 18 μm vertically and the total sensing area is 5.7 mm x 4.4 mm. The resolution is 56 lines/mm vertically and 33 lines/mm horizontally.

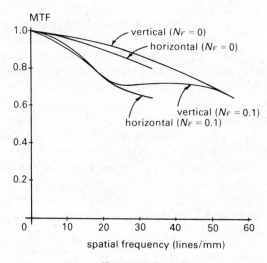

FIGURE 8.16

Theoretical variation of the modulation transfer function of the Fairchild area image sensor with spatial frequency. The array contains 190 horizontal by 244 vertical elements. The parameter $N\epsilon$ indicates the relative loss of charge in N transfers.

Incident photons pass through a transparent polycrystalline silicon gate structure and are absorbed in single crystalline silicon producing free electrons. These are collected if the element is positively biased, and the amount of charge stored is proportional to the incident intensity and the storage time. With no illumination, thermally generated charges will be stored and this limits the sensitivity or dynamic range of the imaging elements. It is found that thermally

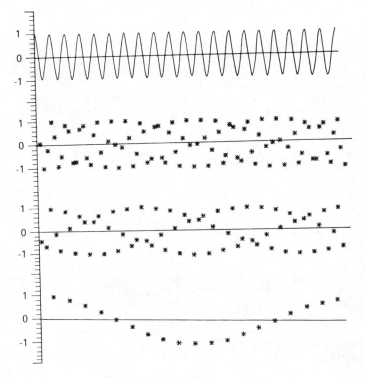

FIGURE 8.17

Aliasing occurs when insufficient samples are taken per cycle of a waveform

generated charge produces around 0.6 mV of video output while strong illumination produces around 200 mV. The entire array can be scanned as rapidly as 100 frames per second, or 19 000 columns per second. Thus 4 636 000 elements are scanned per second giving a video data rate of 4.636 MHz.

Two factors affect the modulation transfer function of the CCD imaging device, the efficiency of transfer of charge from element to element, and the diffusion of charge from one element to another. The variation of the MTF for the 190 x 240 element array is shown in figure 8.16. Although there is a slight degradation at higher spatial frequencies, the MTF of this device is relatively constant up to the limiting values of 56 or 33 lines/mm.

Because of the restricted spatial resolution of the CCD imaging device, any attempt to record an image with higher spatial frequency will result in aliasing (see figure 8.17). The term 'aliasing' is derived from communications to explain the effect of sampling of a signal. Provided the signal is sampled at a rate greater than twice its maximum frequency, no distortion will result. Sampling at a

FIGURE 8.18

This photograph shows a novel application of a charge couple device image sensor. As the carrots pass in front of the image sensor and circuit on the left, the orientation and time of arrival of the carrot is sensed so that the stalk may be cut off. (Courtesy of European Electronic Systems Ltd.)

slower rate results in catastrophic distortion of the signal. In the case of a sinusoidal waveform it is necessary to take at least two samples per cycle in order to avoid distortion. Aliasing in a two-dimensional image will result in loss of detail or may result in the introduction of spurious detail.

8.10   Information Capacity of a CCD Image Sensing Array

The information capacity of an imaging device is determined by the total number of elements (or resolvable cells) in the array, and by the number of levels of intensity which each element can distinguish. For q-level elements in a $K \times l$ array the information capacity is $C = Kl \log_2 q$, as in equation 8.18.

In the example of the 190 x 244 element CCD array, the number of levels of intensity that may be distinguished is dependent on the amount of charge, which accumulates in each element as a result of thermal energy with no illumination. It is estimated that around 5×10^4 electrons are stored in each element under bright illumination, while the rms number of thermally generated electrons per element under dark conditions is 10. Thus approximately 5000 distinct levels of intensity may be distinguished (this may be an over estimate). The information capacity of a 190 x 244 element array is therefore

$$C = 190 \times 244 \times \log_2 (5000) = 569\ 700 \text{ bits}$$

or approximately 22 700 bits/mm^2. Although this is much smaller than the 4 000 000 bits/mm^2 capacity of high resolution film, it is nevertheless a very large information capacity in view of the limited spatial resolution of the CCD array.

The high information capacity of the 190 x 244 element array presents an interesting problem of utilization. How can optical images be coded so as to take advantage of this information capacity? The technique of holography, which makes such efficient utilisation of film, is ruled out since the limited spatial resolution of the CCD array limits the field of view (θ_0) to less than $1°$. A method of coding remains to be discovered which has small spatial frequencies but large variations in intensity.

Appendix A: Photometric Units

A casual reading of manufacturers' handbooks reveals that the light output of various lasers is quoted in watts (W), the conventional measure of electrical power, while the emission of an LED is usually quoted in candelas (cd). The reason for the use of two different sets of units is that the LED is usually marketed for its visual effect as a panel illuminator or digital display, while the laser is intended for optical communication or other applications where its true power output is of importance. The visual effect of a light source has been quantified by measurement and standardisation of the response of the human eye, and defining photometric units in terms of this response. Another difference between the usual measure of power (the watt) and the measure of photometric intensity (the candela), is that the candela is power per unit solid angle.

The basic unit of solid angle is the steradian. For a solid angle having its apex at the centre of a sphere, the measure, in steradians, is the ratio of the area subtended at the surface of the sphere to the square of its radius. The solid angle of a sphere is 4π steradians, and that of a hemisphere is 2π steradians. The steradian is the three-dimensional equivalent of the radian.

The human body can sense electromagnetic radiation in at least two ways: through heating in the body (microwaves or sunlight can warm skin tissue), or through vision. The eye is sensitive to light with a wavelength between about 400 nm and 750 nm under daylight conditions, with a peak response at 550 nm. At night the spectral response of the eye shifts toward the blue, with a peak at about 510 nm.

The relative visibility curve of the eye is a plot of the inverse of the incident power required to give the same sensation of brightness, at each wavelength throughout the visible spectrum. The standard curve for daylight vision is shown in figure A.1. In comparing different sources of light (such as a collection of LEDs), it is customary to compare their luminous intensity, that is, the brightness they would produce in the human eye under normal daylight conditions. The basic unit of luminous intensity is the candela and it can be said that the magnitude of the candela is such that the luminance of a blackbody radiator at the freezing temperature of platinum is 60 candelas per square centimetre. The freezing temperature of platinum is $T = 2042°$ K. A blackbody radiator at that temperature has an optical emission whose spectrum is centred at a wavelength of 1420 nm in the infrared, with a spectral distribution as shown in figure A.2.

In order to calculate the luminance of the blackbody at $2042°$ K it is necessary to multiply its emission spectrum by the relative visibility curve of the eye and integrate the resulting curve over the visible spectrum. The spectral distribution of the luminance of a blackbody at $2042°$ K as seen by the human eye under

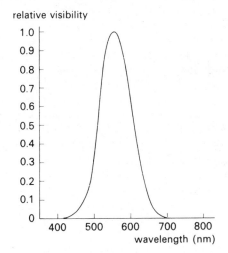

FIGURE A.1
The standard spectral response for daylight vision of the human eye

daylight conditions is shown in figure A.3. The integral of this spectrum is approximately 88.2 mW cm^{-2}sr^{-1}, which is equal to 60 cd/cm^{-2}. Thus 680 cd is equivalent to 1 W/sr, or 680 lumens (680 lm) is equivalent to 1 W in the visible spectrum.

Compared with the spectral response of the eye, the emission from an LED has a narrow spectrum, and may be approximated as a monochromatic source. The brightness of an LED will depend on its colour, because of the relative

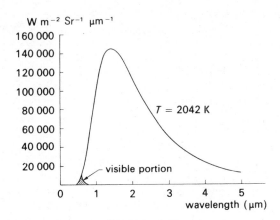

FIGURE A.2
Emission spectrum of a blackbody radiator at 2042° K

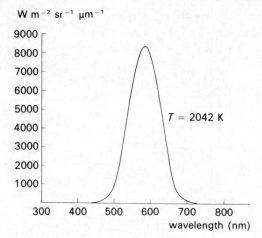

FIGURE A.3

Luminous intensity of a blackbody at 2042° K as seen by the human eye

visibility curve of the eye. This brightness is specified by the luminous efficacy, η_v, of a monochromatic source. The luminous efficacy of an optical source is the ratio of the emitted power in lumens to that in watts at a particular wavelength. For example, an LED which emits red light at 635 nm has a luminous intensity of 24 mcd. The luminous efficacy at this wavelength is 147 lm/W. Thus its radiant intensity is 24/147 or 0.163 mW/sr. The variation of luminous efficacy with wavelength is shown in figure A.4.

It is common practice to specify the luminous intensity of an LED rather than its total luminous power output. The reason is that different devices emit

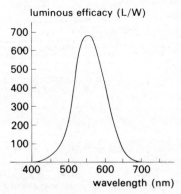

FIGURE A.4

Variation of luminous efficacy with wavelength

their light over a cone of angles ranging from as small as a few degrees, to as large as 90° or more. Obviously an LED which emits a narrow beam of light will have a higher luminous intensity at the centre of the beam for the same luminous power output. The intensity of an optical source may be expressed as

$$I(r, \theta) = P(\theta)/r^2$$

where the far field pattern $P(\theta)$, measured at a large distance r from the source, is the radiant intensity expressed in watts per steradian (or candelas in photometric units). If a photodetector of area A is placed on the axis of the optical source, it would intercept a total optical power equal to

$$I(r, \theta)A = \frac{P(\theta)A}{r^2}$$

Here A/r^2 is the solid angle intercepted by the photodetector.

The total power radiated by the optical source may be calculated from its far field radiation pattern, provided the axial radiant intensity is known. Thus

$$P_T = \int_0^{2\pi} \int_0^{\pi} I(r, \theta)r^2 \sin\theta \ d\theta \ d\Phi = 2\pi \int_0^{\pi} P(\theta) \sin\theta \ d\theta$$

If the axial radiant intensity, $P(0)$, is in watts per steradian, then P_T is in watts, while if it is in candelas P_T is in lumens. The reader may verify that the solid angle of a cone of angular width $2\theta_{1/2}$ is $\Omega(\theta_{1/2}) = 4\pi \sin^2(\theta_{1/2}/2)$.

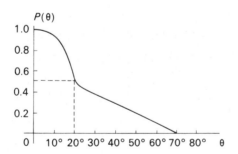

FIGURE A.5
Radiation pattern of an optical source

The photometric units adopted as the international standard are the lumen and the candela. However, several different systems are still in widespread use, the most common term encountered being the footcandle. One footcandle is equal to 1 lm/ft^2, or 10.764 lm/m^2. Listed below are the most commonly needed conversions and other useful information.

680 lm = 1 W   (blackbody source at 2042° K)

1 candela (cd) = 1 lumen (lm)/ steradian (sr)

The lumen is the measure of luminous flux or total optical power.

The candela is the measure of the luminous intensity or optical power per unit solid angle. Typical values of luminous efficacy for LEDs are shown in table A.1

TABLE A.1

Wavelength (nm)	Luminous efficacy (lm/W)
655 (red)	60
635 (red)	135
585 (yellow)	540
565 (green)	640

1 footcandle (lm/ft^2) = 10.76 lx (lm/m^2)

$\qquad\qquad\qquad$ = 0.001076 phot (lm/cm^2)

1 footlambert = 0.001076 lambert

$\qquad\qquad$ = 1.076 millilamberts

$\qquad\qquad$ = 3.426 nits (cd/m^2)

$\qquad\qquad$ = 0.0003426 stilbs (cd/cm^2)

$\qquad\qquad$ = 0.3183 candela/ft^2

$\qquad\qquad$ = 10.76 apostilbs

Appendix B: Physical Constants

Electronic charge
$$e = 1.602 \times 10^{-19} \, c$$
Electron rest mass
$$m_0 = 9.109 \times 10^{-31} \, \text{kg}$$
Planck's constant
$$h = 6.626 \times 10^{-34} \, \text{J s}$$
Boltzmann's constant
$$k = 1.380 \times 10^{-23} \, \text{J/K}$$
Speed of light in vacuum
$$c = 2.998 \times 10^{8} \, \text{m/s}$$
Permittivity constant
$$\epsilon_0 = 8.854 \times 10^{-12} \, \text{F/m}$$
Permeability constant
$$\mu_0 = 4\pi \times 10^{-7} \, \text{H/m}$$

References

Chapter 1 Photons and Matter

Dash, W. C., and Newman, R., 'Intrinsic Optical Absorption in Single-Crystal Germanium and Silicon at 77 K and 300 K', *Phys. Rev.* 99 (1955) 1151.

Hill, D. E., 'Infrared Transmission and Fluorescence of Doped Gallium Arsenide', *Phys. Rev.*, 133 (1964) A866.

Kittel, C., *Introduction to Solid State Physics* (Wiley, New York, 1976).

Scully, M. O., and Sargent, M., 'The Concept of the Photon', *Physics Today*, 25 (1972) 3.

Sze, S. M., *Physics of Semiconductor Devices* (Wiley, New York, 1969).

Chapter 2 The Light Emitting Diode

Hewlett-Packard Optoelectronics Division, *Optoelectronic Applications Manual* (McGraw-Hill, New York, 1977).

Texas Instruments, *Optoelectronics Theory and Practice* (Texas Publications, Dallas, 1977).

Chapter 3 Solid State Photodetectors – The Photoconductor

Bube, R. H., *Photoconductors* (Wiley, New York, 1960).

Keyes, R. J., and Kingston, R. H. 'A Look at Photon Detectors', *Physics Today*, 25 (1972) 3.

Van Der Ziel, A. *Solid State Physical Electronics* (Prentice-Hall, Englewood Cliffs, N.J., 1968).

Chapter 4 Solid State Photodetectors – The Photodiode and Phototransistor

Clayton, G. B. *Operational Amplifiers* (Butterworths, London, 1971).

Di Domenico, M., and Svelto, O. 'Solid State Photodetection – Comparison between Photodiodes and Photoconductors', *Proc. IEEE*, 52 (1964) 136. (See also references for chapter 2 and 3.)

Tobey, G. E., Graeme, J. G., and Huelsman, L. P., *Operational Amplifier – Design and Applications* (McGraw-Hill, New York, 1971).

Chapter 5 Noise in Optoelectronic Devices

Bennett, W. R., *Electrical Noise* (McGraw-Hill, New York, 1960).

Papoulis, A., *Probability, Random Variables and Stochastic Processes* (McGraw-Hill, New York, 1965).

Wozencraft, J. M., and Jacobs, I. M. *Principles of Communication Engineering* (Wiley, New York, 1965).

Chapter 6 The Solar Cell

Backus, C. E., *Solar Cells* (IEEE Press, New York, 1976).

Campbell, I. M., *Energy and the Atmosphere: A Physical–Chemical Approach* (Wiley, London, 1977).

Chapin, D. M., Fuller, C. S., and Pearson, G. L., 'A New Silicon p–n Junction Photocell for Converting Solar Radiation into Electrical Power', *J. appl. Phys.*, 25 (1954) 676.

Prince, M. B. 'Silicon Solar Energy Converters', *J. appl. Phys.*, 26 (1955) 5.

Smits, F. M., 'History of Silicon Solar Cells', *IEEE Trans. electronic Devices*, Ed-23 (1976) 7.

Chapter 7 The Laser Diode

Barnoski, M. K., *Fundamentals of Optical Fibre Communications* (Academic Press, London, 1976).

Boyle, W. S., 'Light-Wave Communications', *Scient. Am.*, 237 (1977) 2.

Midwinter, J. E., *Optical Fibres for Transmission*, (Wiley, New York, 1979).

Panish, M. B., 'Heterostructure Injection Lasers', *Proc. IEEE*, 64 (1976) 10.

Panish, M. B., and Hayashi, I., 'A New Class of Diode Lasers', *Scient. Am.*, 225 (1971) 1.

Stern, J. R., 'Optical Fibre Transmission Systems: Properties of the Fibre Cables Installed for the 8 Mbit/s and 140 Mbit/s Feasibility Trials', POEEJ,71 (1979)

Chapter 8 Optical Imaging Techniques

Boyle, W. S., and Smith, G. E. 'The Inception of Charge-Coupled Devices', *IEEE Trans. electronic Devices*, Ed-23 (1976) 7.

Carlson, F. P., *Introduction to Applied Optics for Engineers* (Academic Press, New York, 1977).

Develis, J. B., and Reynolds, G. O., *Theory and Applications of Holography* (Addison-Wesley, Reading, Mass., 1967).

Sequin, C. H., Zimany, E. J. Jr., Tompsett, M. F., and Fuls, E. N., 'All-Solid-State Camera for the 525-Line Television Format', *IEEE Trans. electronic Devices*, Ed-23 (1976) 2.

Stroke, G. W., *An Introduction to Coherent Optics and Holography* (Academic Press, New York, 1966).

Tompsett, M. F., Amelio, G. F., and Smith, G. E., 'Charge Coupled 8-Bit Shift Registers', *Appl. Phys. Lett.*, 17 (1970) 111.

Yu, F. T. S., *Introduction to Diffraction, Information Processing, and Holography* (MIT Press, New York, 1973).

Yu, F. T. S., 'Markov Photographic Noise', *J. opt. Soc. Am.*, 59 (1969) 3.

Yu, F. T. S., 'Information Channel Capacity of a Photographic Film', *IEEE Trans. Inf. Theory*, IT-16 (1970) 4.

Index